MEASURES ON INFINITE DIMENSIONAL SPACES

SERIES IN PURE MATHEMATICS

Part I. Monographs and Textbooks

Volume 1: Total Mean Curvature and Submanifolds of Finite Type
B Y Chen

Volume 3: Structures on Manifolds
K Yano & M Kon

Volume 4: Goldback Conjecture
Wang Yuan (editor)

Part II. Lecture Notes

Volume 2: A Survey of Trace Forms of Algebraic Number Fields
P E Conner & R Perlis

Volume 5: Measures on Infinite Dimensional Spaces
Y Yamasaki

Series in Pure Mathematics — Volume 5

MEASURES ON INFINITE DIMENSIONAL SPACES

Y Yamasaki

Research Institute for Mathematical Sciences
Kyoto University

World Scientific
Singapore • Philadelphia

Published by

World Scientific Publishing Co. Pte. Ltd.
P. O. Box 128, Farrer Road, Singapore 9128
242, Cherry Street, Philadelphia PA 19106-1906, USA

Library of Congress Cataloging in Publication Data

Yamasaki, Yasuo, 1934-
Lecture Notes on Measures on Infinite Dimensional Spaces.

(Series in pure mathematics; v. 5)
"Dedicated to Professor Hisaaki Yoshizawa on his 60th birthday and to Professor Shizuo Kakutani on his 70th birthday."
1. Measure theory. 2. Spaces, Generalized. I. Title. II. Series.
QA312.Y36 1985 515.4'2 85-9381
ISBN 9971-978-52-0

Printed in Singapore by Singapore National Printers (Pte) Ltd.

INTRODUCTORY NOTE

These notes are based on the lectures given at Yale (1979-81) and at Kyoto (1981-82).

The author wishes to dedicate these lecture notes to Professor Hisaaki Yoshizawa (Kyoto University) and Professor Shizuo Kakutani (Yale University). The author started his study of functional analysis under the direction of Professor Yoshizawa, who also kindly suggested the author's visit to Yale. Professor Kakutani arranged so heartily the author's stay and study at Yale, both stimulating and comfortable for the author.

CONTENTS

PART A: EXTENDABILITY OF A FAMILY OF MEASURES TO A σ-ADDITIVE MEASURE
(Kolmogorov-Bochner-Minlos Theory)

Introduction 3

Chapter 1. Preliminary discussions 4
§1. Explanation of the problem 4
§2. Tychonov's theorem 6
§3. Hopf's theorem 9

Chapter 2. Direct product and projective limit 14
§4. Measurable spaces, their product and limit 14
§5. Extension problems and counter examples 17
§6. Extension theorem for direct product 22
§7. Extension theorem for projective limit 26
§8. Non-countable direct product and projective limit 30
§9. Compact regular measurable space 35
§10. Borel field 38
§11. Baire field 42
§12. Product measure 46
§13. Suslin set, Luzin set 50
§14. Standard measurable space 56

Chapter 3. Measures on vector spaces 63
§15. Explanation of the problem 63
§16. Relation with Bochner's theorem 68
§17. Minlos' theorem 72
§18. Sazonov topology 77
§19. Supplementary results 81
§20. Nuclear space 85
§21. Heredity 89
§22. Dual space 95

Appendix. Definition of nuclearity without using Hilbertian semi-norms 98

Note 106

PART B: INVARIANCE AND QUASI-INVARIANCE OF MEASURES ON INFINITE DIMENSIONAL SPACES

Introduction 111

Chapter 1. Invariant measure on a group 113
§1. Measurable group, invariant and quasi-invariant measure 113

§2. Haar measure on a locally compact group 118
§3. Haar measure on a thick group 125
§4. Weil topology 133
§5. Case of a vector space 138

Chapter 2. Gaussian measures and related problems 144
§6. Quasi-invariance and ergodicity 144
§7. Absolute continuity of projective limit measures 147
§8. Gaussian measures 151
§9. E′-quasi-invariance and E′-ergodicity 154
§10. Mutual equivalence 160
§11. Rotationally invariant measures 163
§12. Representation of $L^2(\mu)$ 167

Chapter 3. The set Y_μ of all quasi-invariant translations 174
§13. Convolution of measures 174
§14. Linearization of a topology on a vector space 175
§15. Characteristic topology 177
§16. Evaluation of Y_μ in terms of τ_μ 181
§17. Some applications 184
§18. Kakutani topology 189
§19. Evaluation of Y_μ in terms of Kakutani topology 195

Chapter 4. Product measures on $\mathbb{R}^\infty$ 202
§20. Product of one-dimensional probability measures 202
§21. Stationary product measures 208
§22. Gaussian measures and stationary products 212
§23. Estimation of Y_μ and R_μ 216
§24. Non-stationary product measures 219
§25. (ℓ^2)-quasi-invariance 222

Chapter 5. $\mathbb{R}_0^\infty$-invariant measures on $\mathbb{R}^\infty$ 227
§26. Infinite dimensional Lebesgue measure 227
§27. $\mathbb{R}_0^\infty$-ergodicity and mutual equivalence 230
§28. Equivalent probability measure of product type 233
§29. The converse problem of §28 235
§30. Linear transformation of μ_∞ 240
§31. Rotational invariance and ergodicity 244
§32. Invariance under homotheties 247

Note 251

Index 255

MEASURES ON INFINITE DIMENSIONAL SPACES

Part A

Extendability of a Family of Measures to a σ-Additive Measure

(Kolmogorov-Bochner-Minlos theory)

PART A

Extendability of a family of measures to a σ-additive measure (Kolmogorov-Bochner-Minlos theory)

Introduction

The extension theorem provides a way to construct an infinite dimensional measure as the limit of finite dimensional ones. It was first established by Kolmogorov for the case of $\mathbb{R}^\infty$, direct product of real lines. Later, his classical result was generalized to the case of: 1°) direct product of σ-compact metric spaces, 2°) projective limit (instead of direct product), and 3°) complete separable metric space (instead of σ-compact one). Among them, 1°) seems to be most fundamental. In this lecture note, the author will prove 1°) under a little weaker condition, and derive the results of 2°) and 3°) from the modified 1°).

In many occasions, we need the extension theorem on some function spaces or distribution spaces, which are not given as direct product nor projective limit. The extendability of measures on an infinite dimensional vector space E is often discussed in connection with the continuity of the characteristic functions on E', the dual of E. A satisfactory theory has been established for Hilbert spaces, and generalized to Hilbert type spaces, then to nuclear spaces. Any discussion without using inner product is very difficult, and we have no useful result along this line.

CHAPTER 1. PRELIMINARY DISCUSSIONS

§1. Explanation of the problem

In this section we shall consider examples illuminating our problem.

Let $\mathbb{R}^\infty$ be the set of all real sequences:

(1.1) $\mathbb{R}^\infty = \{(x_1, x_2, \cdots) \mid \text{each } x_k \text{ is a real number}\}$,

and let p_n be the projection from $\mathbb{R}^\infty$ to $\mathbb{R}^n$:

(1.2) $p_n : (x_1, x_2, \cdots) \to (x_1, x_2, \cdots, x_n)$.

For a given measure μ on $\mathbb{R}^\infty$, its marginal measure μ_n is defined on $\mathbb{R}^n$ as follows:

(1.3) $\mu_n(E) = \mu(p_n^{-1}(E)), \quad \forall E \subset \mathbb{R}^n$.

The family $\{\mu_n\}$ satisfies evidently

(1.4) $\mu_n(p_{mn}^{-1}(E)) = \mu_m(E), \quad \forall E \subset \mathbb{R}^m, \quad n > m$,

where p_{mn} is the projection from $\mathbb{R}^n$ to $\mathbb{R}^m$ similar as (1.2). The condition (1.4) is called the self-consistency condition for $\{\mu_n\}$.

The converse is our problem. Namely, whenever a self-consistent sequence $\{\mu_n\}$ is given, does there exist a σ-additive measure μ which satisfies (1.3)? This question was answered affirmatively by Kolmogorov and became the starting point of the theory of extendability. ($\{\mu_n\}$ is said to be extended to μ if (1.3) holds, and $\{\mu_n\}$ is said to be extendable if such a μ exists).

The case of $\mathbb{R}^{[0,1]}$ gives an apparently different example.

For a measure μ on $\mathbb{R}^{[0,1]}$, the set of all real functions on $[0,1]$, its marginal measure $\mu_{t_1,t_2,\cdots,t_n}$ is defined on $\mathbb{R}^n$;

(1.5) $\quad \mu_{t_1,t_2,\cdots,t_n}(E) = \mu(p^{-1}_{t_1,t_2,\cdots,t_n}(E)), \quad \forall E \subset \mathbb{R}^n,$

where $p_{t_1,t_2,\cdots,t_n}$ maps $f(t) \in \mathbb{R}^{[0,1]}$ to $(f(t_1), f(t_2), \cdots, f(t_n)) \in \mathbb{R}^n$. The family $\{\mu_{t_1,t_2,\cdots,t_n}\}$ satisfies the self-consistency condition similar as (1.4).

The converse is also our problem, and answered affirmatively. Namely whenever $\{\mu_{t_1,\cdots,t_n}\}$ satisfies the self-consistency condition, the corresponding measure μ exists on $\mathbb{R}^{[0,1]}$. This is an important way to define a measure on $\mathbb{R}^{[0,1]}$ (or on $\mathbb{R}^\infty$ in the first example).

Actually the second example can be reduced to the first one. Extendability does not depend on the cardinal number of direct product, and the problem can always be reduced to the case of countable direct product.

On the contrary, the σ-compactness of $\mathbb{R}$ is essential. The extension theorem can be proved for the direct product of σ-compact metric spaces $\Omega^{(k)}$. In this case, μ_n is a measure on $\Omega_n = \prod_{k=1}^{n} \Omega^{(k)}$, and the self-consistency condition is given by (1.4) where p_{mn} is the projection from Ω_n to Ω_m. Such a family $\{\mu_n\}$ can be extended to a measure on $\Omega = \prod_{k=1}^{\infty} \Omega^{(k)}$.

One more generalization is the introduction of projective limit (due to Bochner). Suppose that Ω_n and some surjective mapping p_{mn} is given with the condition $p_{mn} = p_{m\ell} \circ p_{\ell n}$ $(m<\ell<n)$, and assume that the family $\{\mu_n\}$ satisfies (1.4). In this

case, we should construct Ω, the projective limit of $\{\Omega_n\}$, before we discuss the extendability of $\{\mu_n\}$. Using this set up, for instance we can consider the projective limit of uniform measures on n-dimensional spheres. Let Ω_n be the n-dimensional unit sphere. Ω_{n-1} can be regarded as the equator of Ω_n. For $P \in \Omega_n$, we define $p_{n-1,n}(P)$ as the intersection of the equator and the meridian through P. This is an example essentially different from direct product.

In this lecture, every measure is assumed to be finite. However the obtained result is valid even for σ-finite measures. Extension theorem for non-σ-finite measures is more complicated, and we do not treat it here.

In the next chapter, we shall formulate the extension theorem in an exacter way, and prove it. For the proof, Tychonov's theorem and Hopf's theorem are essentially used, so that in the rest of this chapter, we shall explain these two theorems.

§2. Tychonov's theorem

In this lecture, every topological space is assumed to be Hausdorff unless mentioned otherwise.

Definition 2.1. A topological space X is said to be compact if

(C) Every open covering of X containes a finite covering.

The condition (C) is equivalently rewritten as:

(C1) For a family of closed sets, the finite intersection property implies the complete intersection property,

and also as:

(C2) For a family of sets, the finite intersection property implies the complete intersection property for closures.

This means that for a family of sets $\mathcal{E}$, the following (2.1) implies (2.2).

$$\bigcap_{E\in\mathcal{F}} E \neq \emptyset \quad \text{for every finite family } \mathcal{F}\subset\mathcal{E}, \tag{2.1}$$

$$\bigcap_{E\in\mathcal{E}} \bar{E} \neq \emptyset. \tag{2.2}$$

<u>Definition 2.1'.</u> A subset C of a topological space X is said to be compact if C is compact in the topology of subspace of X.

<u>Definition 2.2.</u> For a family of topological spaces $\{X_\alpha\}_{\alpha\in A}$, consider the direct product set $X=\prod_{\alpha\in A} X_\alpha$. The product topology (or weak topology) on X is defined as the weakest topology in which every projection $p_\alpha\colon X \to X_\alpha$ is continuous.

The fundamental system of neighbourhoods in the product topology is given by

$$U = \bigcap_{k=1}^{n} p_{\alpha_k}^{-1}(U_{\alpha_k}), \tag{2.3}$$

where $\{\alpha_1,\alpha_2,\cdots,\alpha_n\}$ runs over all finite subsets of A and U_{α_k} is a neighbourhood in X_{α_k}.

<u>Theorem 2.1.</u> (Tychonov) If all X_α is compact, then $X=\prod_{\alpha\in A} X_\alpha$ is compact in the product topology.

<u>Corollary.</u> If C_α is a compact set in a topological space X_α, then $C=\prod_{\alpha\in A} C_\alpha$ is compact in the topological product space $X = \prod_{\alpha\in A} X_\alpha$.

Proof (of Th.). Let $\mathcal{E}$ be a family of subsets of X satisfying finite intersection property (=FIP), and we shall prove (2.2). By Zorn's lemma, there exists a maximal family satisfying FIP and containing $\mathcal{E}$. (Since FIP is kept valid for the union of increasing families, Zorn's lemma is applicable). So that it is sufficient to prove (2.2) under the assumption that $\mathcal{E}$ is a maximal family.

If $\mathcal{E}$ is a maximal family with FIP, it is easily shown that

1) $E_1, E_2 \in \mathcal{E} \Longrightarrow E_1 \cap E_2 \in \mathcal{E}$

2) $A \subset X,\ A \cap E \neq \emptyset$ for $\forall E \in \mathcal{E} \Rightarrow A \in \mathcal{E}$

On the other hand, FIP for $\mathcal{E}$ implies FIP for $p_\alpha(\mathcal{E}) = \{p_\alpha(E) \mid E \in \mathcal{E}\}$. Since each X_α is compact, we have $\bigcap_{E \in \mathcal{E}} \overline{p_\alpha(E)} \neq \emptyset$. This means

$$(2.4) \qquad \forall \alpha,\ \exists x_\alpha \in X_\alpha,\ \forall U_\alpha (\text{neigh. of } x_\alpha)\ \forall E \in \mathcal{E}$$

$$U_\alpha \cap p_\alpha(E) \neq \emptyset.$$

But the second line of (2.4) is equivalent with $p_\alpha^{-1}(U_\alpha) \cap E \neq \emptyset$, therefore from 2) we get $p_\alpha^{-1}(U_\alpha) \in \mathcal{E}$. Thus from 1) we get $\bigcap_{k=1}^{n} p_{\alpha_k}^{-1}(U_{\alpha_k}) \in \mathcal{E}$, so that

$$(2.5) \qquad \forall E \in \mathcal{E},\ \bigcap_{k=1}^{n} p_{\alpha_k}^{-1}(U_{\alpha_k}) \cap E \neq \emptyset.$$

Comparing with (2.3), this means that $x = (x_\alpha)$ belongs to $\bigcap_{E \in \mathcal{E}} \overline{E}$.

(q.e.d.)

§3. Hopf's theorem

Definition 3.1. Let X be a set. A family of subsets of X is called a σ-algebra (resp. finite algebra), if it is closed under complement and countable (resp. finite) union.

In this lecture, a σ-algebra (resp. finite algebra) is denoted with $\mathcal{B}$ (resp. $\mathcal{F}$), unless mentioned otherwise.

Definition 3.2. A non-negative function μ on $\mathcal{B}$(resp. $\mathcal{F}$) is called a measure if it is countably (resp. finitely) additive, namely if it satisfies

$$(3.1) \qquad E_i \cap E_j = \emptyset \text{ for } i \neq j \implies \mu(\bigcup_i E_i) = \sum_i \mu(E_i),$$

where $\bigcup_i$ is a countable (resp. finite) union.

In this lecture, a measure is always assumed to be finite.

Definition 3.3. A non-negative function on 2^X (the family of all subsets of X) is called a Caratheodory's outer measure if it is monotonic and countably subadditive, namely if it satisfies

$$(3.2) \qquad E_1 \subset E_2 \implies \mu(E_1) \leq \mu(E_2)$$

$$(3.3) \qquad \mu(\bigcup_i E_i) \leq \sum_i \mu(E_i).$$

A set E is said to be μ-measurable if

$$(3.4) \qquad \forall Y \subset X,\ \mu(Y) = \mu(Y \cap E) + \mu(Y \cap E^c).$$

Theorem 3.1. For a Caratheodory's outer measure μ, the family of all μ-measurable sets is a σ-algebra, and the restriction of μ on this σ-algebra is a measure.

Proof. Let $\mathcal{B}$ be the family of all μ-measurable sets. Evidently $\mathcal{B}$ is closed under complement; $E \in \mathcal{B} \Longleftrightarrow E^c \in \mathcal{B}$. Suppose that E_i $(1 \le i < \infty)$ belongs to $\mathcal{B}$. Then we have

$$\mu(Y) = \mu(Y \cap E_1) + \mu(Y \cap E_1^c) = \mu(Y \cap E_1) + \mu(Y \cap E_1^c \cap E_2) + \mu(Y \cap E_1^c \cap E_2^c)$$

$$= \sum_{i=1}^{n} \mu(Y \cap E_1^c \cap \cdots \cap E_{i-1}^c \cap E_i) + \mu(Y \cap E_1^c \cap \cdots \cap E_n^c).$$

The last term is larger than $\mu(Y \cap (\bigcup_{i=1}^{\infty} E_i)^c)$, so in the limit of $n \to \infty$,

$$\mu(Y) \ge \sum_{i=1}^{\infty} \mu(Y \cap E_1^c \cap \cdots \cap E_{i-1}^c \cap E_i) + \mu(Y \cap (\bigcup_{i=1}^{\infty} E_i)^c). \tag{3.5}$$

The first term is larger than $\mu(Y \cap (\bigcup_{i=1}^{\infty} E_i))$ from (3.3), so that we have

$$\mu(Y) \ge \mu(Y \cap (\bigcup_{i=1}^{\infty} E_i)) + \mu(Y \cap (\bigcup_{i=1}^{\infty} E_i)^c). \tag{3.6}$$

This implies $\bigcup_{i=1}^{\infty} E_i \in \mathcal{B}$. Especially if we choose as $Y = \bigcup_{i=1}^{\infty} E_i$ and if we assume $\{E_i\}$ is mutually disjoint, then (3.5) shows that

$$\mu(\bigcup_{i=1}^{\infty} E_i) \ge \sum_{i=1}^{\infty} \mu(E_i).$$

This means that μ is a measure on $\mathcal{B}$, because the converse inequality is true from (3.3). (q.e.d.)

Theorem 3.2 (Hopf). Let $\mathcal{B}(\mathcal{F})$ be the smallest σ-algebra containing $\mathcal{F}$. A finitely additive measure μ on $\mathcal{F}$ can be extended to a σ-additive measure on $\mathcal{B}(\mathcal{F})$, if and only if

(H) $\quad F_i \in \mathfrak{F}\ ;\ F_1 \supset F_2 \supset \cdots \supset F_n \supset \cdots$

$$\lim_{n\to\infty} \mu(F_n) > 0 \Longrightarrow \bigcap_{n=1}^{\infty} F_n \neq \emptyset .$$

Proof. The condition (H) is necessary, because if μ is extended to a σ-additive measure ν on $\mathcal{B}(\mathfrak{F})$, we have $\lim_{n\to\infty} \mu(F_n) = \nu(\bigcap_{n=1}^{\infty} F_n)$.

Conversely, assume the condition (H). It is easily checked that

$$\mu^*(E) = \inf\{ \sum_{i=1}^{\infty} \mu(F_i) \mid E \subset \bigcup_{i=1}^{\infty} F_i \} \tag{3.7}$$

becomes a Caratheodory's outer measure. We shall prove that every $F \in \mathfrak{F}$ is μ^*-measurable and that $\mu^* = \mu$ on $\mathfrak{F}$.

Suppose that $Y \subset X$ and $\bigcup_{i=1}^{\infty} F_i \supset Y$ are given. Since μ is finitely additive, we have

$$\sum_{i=1}^{\infty} \mu(F_i) = \sum_{i=1}^{\infty} (\mu(F_i \cap F) + \mu(F_i \cap F^c)) \geq \mu^*(Y \cap F) + \mu^*(Y \cap F^c).$$

Thus, taking the infimum of the left side, we have

$$\mu^*(Y) \geq \mu^*(Y \cap F) + \mu^*(Y \cap F^c).$$

This means F is μ^*-measurable.

Next, from (3.7) $\mu^*(F) \leq \mu(F)$ is evident. We shall prove $\mu^*(F) \geq \mu(F)$, namely prove

$$F \subset \bigcup_{i=1}^{\infty} F_i \Rightarrow \mu(F) \leq \sum_{i=1}^{\infty} \mu(F_i). \tag{3.8}$$

Under the assumption of (3.8), if we put $F'_n = \bigcap_{i=1}^{n} F_i^c \cap F$, then $\{F'_n\}$ is decreasing and $\bigcap_{n=1}^{\infty} F'_n = \emptyset$. From (H), we have $\mu(F'_n) \to 0$, therefore $\forall \varepsilon > 0$, $\exists n$, $\mu(\bigcap_{i=1}^{n} F_i^c \cap F) < \varepsilon$, thus we get

$$\mu(F) - \varepsilon < \mu(\bigcup_{i=1}^{n} F_i \cap F) \leq \sum_{i=1}^{n} \mu(F_i \cap F) \leq \sum_{i=1}^{\infty} \mu(F_i).$$

This holds for every $\varepsilon > 0$, so that we have (3.8). (q.e.d.)

For a given measure μ on $\mathcal{B}$ (resp. $\mathfrak{F}$), define a metric d as follows:

$$(3.9) \qquad d(E_1, E_2) = \mu(E_1 \Delta E_2),$$

where $E_1 \Delta E_2 = (E_1 \cap E_2^c) \cup (E_1^c \cap E_2)$ is the symmetric difference. d satisfies the following properties.

(1) $d(E_1, E_2) = d(E_2, E_1)$

(2) $d(E_1, E_2) \leq d(E_1, E_3) + d(E_2, E_3)$.

Thus d is a metric on $\mathcal{B}$ (resp. $\mathfrak{F}$). (Rigorously speaking, d is a metric after identifying two sets E_1 and E_2 such that $d(E_1, E_2) = 0$, namely such that $E_1 \Delta E_2$ is a null set.)

(3) $|\mu(E_1) - \mu(E_2)| \leq d(E_1, E_2)$

(4) $d(E_1, E_2) = d(E_1^c, E_2^c)$

(5) $d(\bigcup_i E_{1i}, \bigcup_i E_{2i}) \leq \sum_i d(E_{1i}, E_{2i})$,

where $\bigcup_i$ means a countable (resp. finite) union.

<u>Theorem 3.3</u>. $\mathfrak{F}$ is dense in $\mathcal{B}(\mathfrak{F})$ for any σ-additive measure on $\mathcal{B}(\mathfrak{F})$.

Proof. We shall prove that $\overline{\mathfrak{F}}$ (closure of $\mathfrak{F}$) is a σ-algebra. From (4), $\overline{\mathfrak{F}}$ is closed under complement. Suppose $E_i \in \overline{\mathfrak{F}}$ for $1 \leq i < \infty$, then the inequality

$$d\left(\bigcup_{i=1}^{\infty} E_i, \bigcup_{i=1}^{n} F_i\right) \leq d\left(\bigcup_{i=1}^{\infty} E_i, \bigcup_{i=1}^{n} E_i\right) + \sum_{i=1}^{n} d(E_i, F_i)$$

shows that $\bigcup_{i=1}^{\infty} E_i \in \overline{\mathfrak{F}}$. (q.e.d.)

Theorem 3.4. A σ-additive extension of a finitely additive measure μ on $\mathfrak{F}$ is unique on $\mathfrak{B}(\mathfrak{F})$, if possible.

Proof. Suppose that μ_1 and μ_2 are two extensions. If we consider $\mu_3 = \mu_1 + \mu_2$, then both μ_1 and μ_2 are continuous with respect to the corresponding metric d_3 in virtue of (3) and $d_3 \geq d_1, d_2$. Therefore we have $\mu_1 = \mu_2$ on $\overline{\mathfrak{F}} = \mathfrak{B}(\mathfrak{F})$.

(q.e.d.)

CHAPTER 2. DIRECT PRODUCT AND PROJECTIVE LIMIT

§4, Measurable spaces, their product and limit

Definition 4.1. Let X be a set, and $\mathcal{B}$ be a σ-algebra of subsets of X. $(X,\mathcal{B})$ is called a measurable space, and every $B\in\mathcal{B}$ is called a $\mathcal{B}$-measurable set.

Definition 4.2. Let $(X_1,\mathcal{B}_1)$ and $(X_2,\mathcal{B}_2)$ are two measurable spaces. A map φ from X_1 to X_2 is said to be measurable (with respect to $\mathcal{B}_1$ and $\mathcal{B}_2$) if $\mathcal{B}_1\supset\varphi^{-1}(\mathcal{B}_2)=\{\varphi^{-1}(B_2)\,|\,B_2\in\mathcal{B}_2\}$.

A bijective map φ is called a measurable isomorphism if both φ and φ^{-1} are measurable, namely if $\mathcal{B}_1=\varphi^{-1}(\mathcal{B}_2)$. Two measurable spaces are said to be isomorphic if some measurable isomorphism exists between them.

When $(X_2,\mathcal{B}_2)$ and a map φ from X_1 to X_2 are given, $\varphi^{-1}(\mathcal{B}_2)$ is the smallest σ-algebra with respect to which φ is measurable. When $(X_1,\mathcal{B}_1)$ and φ are given, $\mathcal{B}_1^{\varphi}=\{E\subset X_2\,|\,\varphi^{-1}(E)\in\mathcal{B}_1\}$ is the largest σ-algebra with respect to which φ is measurable.

Let $\mathfrak{A}$ be a family of subsets of X. We shall denote the smallest σ-algebra containing $\mathfrak{A}$ with $\mathcal{B}(\mathfrak{A})$.

Let φ be a map from X_1 to X_2, then we have

$$(4.1)\qquad \varphi^{-1}\Big(\bigcup_{\lambda}\mathfrak{A}_{\lambda}\Big)=\bigcup_{\lambda}\varphi^{-1}(\mathfrak{A}_{\lambda}),$$

$$(4.2)\qquad \varphi^{-1}(\mathcal{B}(\mathfrak{A}))=\mathcal{B}(\varphi^{-1}(\mathfrak{A})).$$

(4.2) can be proved as follows: $\varphi^{-1}(\mathcal{B}(\mathfrak{A}))\supset\varphi^{-1}(\mathfrak{A})$ implies $\varphi^{-1}(\mathcal{B}(\mathfrak{A}))\supset\mathcal{B}(\varphi^{-1}(\mathfrak{A}))$, while $\mathcal{B}(\varphi^{-1}(\mathfrak{A}))^{\varphi}\supset\mathfrak{A}$ implies $\mathcal{B}(\varphi^{-1}(\mathfrak{A}))^{\varphi}\supset\mathcal{B}(\mathfrak{A})$ therefore $\mathcal{B}(\varphi^{-1}(\mathfrak{A}))\supset\varphi^{-1}(\mathcal{B}(\mathfrak{A}))$.

Let Y be a subset of X. The injection $\varphi:Y\to X$ is a special case, and we have

(4.1)' $\quad \bigcup_\lambda \mathfrak{A}_\lambda \cap Y = \bigcup_\lambda (\mathfrak{A}_\lambda \cap Y)$,

(4.2)' $\quad \mathcal{B}(\mathfrak{A}) \cap Y = \mathcal{B}(\mathfrak{A} \cap Y)$.

Let φ be a map from X_1 to X_2, and ψ be a map from X_2 to X_3. Then we have

(4.3) $\quad (\psi \circ \varphi)^{-1}(\mathfrak{A}) = \varphi^{-1}(\psi^{-1}(\mathfrak{A}))$,

because for each $A \in \mathfrak{A}$ we have $(\psi \circ \varphi)^{-1}(A) = \varphi^{-1}(\psi^{-1}(A))$.

<u>Definition 4.3.</u> Let $\{(X^{(k)}, \mathcal{B}^{(k)})\}$ be a sequence of measurable spaces. The product measurable space $(X, \mathcal{B})$ is defined as follows: $X = \prod_{k=1}^{\infty} X^{(k)}$, and $\mathcal{B}$ is the smallest σ-algebra in which all projections $p^{(k)}: X \to X^{(k)}$ are measurable, namely

(4.4) $\quad \mathcal{B} = \mathcal{B}(\bigcup_{k=1}^{\infty} p^{(k)-1}(\mathcal{B}^{(k)}))$.

Similarly, the partial product $(X_n, \mathcal{B}_n)$ is defined as follows: $X_n = \prod_{k=1}^{n} X^{(k)}$ and $\mathcal{B}_n = \mathcal{B}(\bigcup_{k=1}^{n} p_n^{(k)-1}(\mathcal{B}^{(k)}))$, where $p_n^{(k)}$ is the projection from X_n to $X^{(k)}$. Denoting the projection from X to X_n with p_n, we have evidently $p^{(k)} = p_n^{(k)} \circ p_n$, hence $p^{(k)-1}(\mathcal{B}^{(k)}) = p_n^{-1}(p_n^{(k)-1}(\mathcal{B}^{(k)}))$. From this, using (4.1) and (4.2) we get

(4.5) $\quad p_n^{-1}(\mathcal{B}_n) = \mathcal{B}(\bigcup_{k=1}^{n} p^{(k)-1}(\mathcal{B}^{(k)}))$.

Thus, $\{p_n^{-1}(\mathcal{B}_n)\}$ is an increasing family so that $\mathcal{F} = \bigcup_{n=1}^{\infty} p_n^{-1}(\mathcal{B}_n)$ is a finite algebra. Meanwhile from (4.5) we see that $\bigcup_{n=1}^{\infty} p^{(n)-1}(\mathcal{B}^{(n)}) \subset \mathcal{F} \subset \mathcal{B}$ hence we get:

(4.6) $$\mathcal{B} = \mathcal{B}(\mathcal{F}), \qquad \mathcal{F} = \bigcup_{n=1}^{\infty} p_n^{-1}(\mathcal{B}_n).$$

<u>Definition 4.4.</u> Let $\{(X_n, \mathcal{B}_n)\}$ be a sequence of measurable spaces and for $n>m$ let p_{mn} be a measurable surjective map from X_n to X_m with the condition $p_{mn} = p_{m\ell} \circ p_{\ell n}$ for $n>\ell>m$. Then, $\{(X_n, \mathcal{B}_n), p_{mn}\}$ is called a projective sequence. Let $(X, \mathcal{B})$ be the product measurable space of $\{(X_n, \mathcal{B}_n)\}$ and let X_∞ be the following subset of X:

(4.7) $$X_\infty = \{(x_n) \in X \mid p_{mn}(x_n) = x_m \text{ for every } n>m\}.$$

Putting $\mathcal{B}_\infty = \mathcal{B} \cap X_\infty$, we define the projective limit measurable space $(X_\infty, \mathcal{B}_\infty)$.

Let π_n be the projection from $X = \prod_{n=1}^{\infty} X_n$ to X_n, and p_n be the restriction of π_n on X_∞. The map p_n is surjective because for a given $x_n \in X_n$, if we define $x_m = p_{mn}(x_n)$ for $m<n$ and $x_\ell = p_{\ell,\ell+1}(x_{\ell+1})$ for $\ell \geqq n$, the obtained sequence belongs to X_∞ and is mapped to x_n by p_n. From definition (4.7), we have $p_m = p_{mn} \circ p_n$ for $m<n$, so that $p_m^{-1}(\mathcal{B}_m) = p_n^{-1}(p_{mn}^{-1}(\mathcal{B}_m)) \subset p_n^{-1}(\mathcal{B}_n)$ in virtue of the measurability of p_{mn}. Thus, $\{p_n^{-1}(\mathcal{B}_n)\}$ is an increasing family and $\mathcal{F} = \bigcup_{n=1}^{\infty} p_n^{-1}(\mathcal{B}_n)$ is a finite algebra. Meanwhile from (4.1)' and (4.2)', we have $\mathcal{B}_\infty = \mathcal{B} \cap X_\infty = \mathcal{B}(\bigcup_{n=1}^{\infty} \pi_n^{-1}(\mathcal{B}_n) \cap X_\infty) = \mathcal{B}(\bigcup_{n=1}^{\infty} p_n^{-1}(\mathcal{B}_n))$, thus in this case also we have:

(4.6)' $$\mathcal{B}_\infty = \mathcal{B}(\mathcal{F}), \qquad \mathcal{F} = \bigcup_{n=1}^{\infty} p_n^{-1}(\mathcal{B}_n).$$

Finally we shall remark that the direct product is a

special case of the projective limit. Meaning of this statement is as follows: let $\{(X^{(k)},\mathcal{B}^{(k)})\}$ be a sequence of measurable spaces, and $\{(X_n,\mathcal{B}_n)\}$ be the sequence of the partial products. Denoting the projection from X_n to X_m with p_{mn}, $\{(X_n,\mathcal{B}_n),\ p_{mn}\}$ becomes a projective sequence. Then, the projective limit measurable space $(X_\infty,\mathcal{B}_\infty)$ is isomorphic with the infinite product measurable space $(X,\mathcal{B})$ of $\{(X^{(k)}, \mathcal{B}^{(k)})\}$. The measurable isomorphism φ is given by:

$$(4.8) \qquad \varphi : (x^{(1)}, x^{(2)},\cdots, x^{(k)},\cdots) \in X \longleftrightarrow (x_1, x_2,\cdots, x_n,\cdots) \in X_\infty,$$

where $x_n = (x^{(1)},x^{(2)},\cdots,x^{(n)})$.

In this sense, the projective limit is a more generalized setting than the direct product.

§5. Extension problems and counter examples

Definition 5.1. Let $(X,\mathcal{B})$ be a measurable space, and μ be a measure on $\mathcal{B}$. The triplet $(X,\mathcal{B},\mu)$ is called a measure space. Two measure spaces $(X_1,\mathcal{B}_1,\mu_1)$ and $(X_2,\mathcal{B}_2,\mu_2)$ is said to be isomorphic if a measure-preserving measurable isomorphism exists between them.

Definition 5.2. Let $(X,\mathcal{B},\mu)$ be a measure space, $(X',\mathcal{B}')$ be a measurable space, and φ be a measurable map from X to X'. The image measure μ_φ(or $\varphi\circ\mu$) is defined on $\mathcal{B}'$ as follows:

$$(5.1) \qquad \mu_\varphi(B') = \mu(\varphi^{-1}(B')), \quad \forall B' \in \mathcal{B}'.$$

Evidently we have

(5.2) $\mu_{\varphi\circ\psi} = (\mu_\psi)_\varphi$ (or $(\varphi\circ\psi)\circ\mu = \varphi\circ(\psi\circ\mu)$).

The injection $Y\subset X$ is a special case. For a measure μ on $(Y,\mathcal{B}\cap Y)$, we define the imbedded measure ν on $(X,\mathcal{B})$ as follows:

(5.3) $\nu(B) = \mu(B\cap Y), \quad {}^\forall B\in\mathcal{B}.$

We often use the same symbol μ instead of ν. In other words, a measure on $(Y,\mathcal{B}\cap Y)$ can be regarded as a measure on $(X,\mathcal{B})$ by the imbedding.

Definition 5.3. Let $(X,\mathcal{B},\nu)$ be a measure space. A subset $Y\subset X$ is said to be thick (with respect to ν) if

(5.4) $B\in\mathcal{B},\ B\cap Y = \emptyset \Longrightarrow \nu(B)=0.$

For a thick subset Y, there exists a unique measure μ on $(Y,\mathcal{B}\cap Y)$ which satisfies (5.3). ($B_1\cap Y=B_2\cap Y$ implies $(B_1\Delta B_2)\cap Y=\emptyset$, hence $\nu(B_1)=\nu(B_2)$, therefore (5.3) can define μ consistently). This μ is called the trace of ν on Y.

A $\mathcal{B}$-measurable set Y is thick if and only if $\nu(Y^c)=0$. In this case, we have $\mathcal{B}\cap Y\subset\mathcal{B}$, and the trace is just equal with the restriction of ν on $\mathcal{B}\cap Y$.

Definition 5.4. Let $\{(X^{(k)},\mathcal{B}^{(k)})\}$ be a sequence of measurable spaces, and let $\{(X_n,\mathcal{B}_n)\}$ be the sequence of the partial products. Let μ_n be a measure on $\mathcal{B}_n$, and let p_{mn} be the projection from X_n to X_m for $m<n$. The sequence $\{\mu_n\}$ is said to be self-consistent if

(5.5) $p_{mn}\circ\mu_n = \mu_m$ for $m<n$.

Definition 5.4'. Let $\{(X_n,\mathcal{B}_n), p_{mn}\}$ be a projective sequence of measurable spaces. The self-consistency of $\{\mu_n\}$ is defined also by (5.5).

Let μ be a measure on the infinite product or projective limit measurable space. Putting $\mu_n=p_n\circ\mu$, we see from (5.2) that $\{\mu_n\}$ is a self-consistent sequence. The converse is our problem. Namely:

Problem A. Let $\{\mu_n\}$ be a self-consistent sequence of measures on $(X_n,\mathcal{B}_n)$. Does there exist a measure μ on the infinite product or projective limit measurable space such that

$$(5.6) \qquad \forall n, \quad \mu_n = p_n\circ\mu$$

We shall remark that putting $\mu(p_n^{-1}(E))=\mu_n(E)$, we can define consistently a finitely additive measure μ on $\mathcal{F}=\bigcup_{n=1}^{\infty} p_n^{-1}(\mathcal{B}_n)$, because $p_n^{-1}(E_n)=p_m^{-1}(E_m)$ implies $E_n=p_{mn}^{-1}(E_m)$ therefore $\mu_n(E_n)=\mu_m(E_m)$. This μ is σ-additive on each $p_n^{-1}(\mathcal{B}_n)$. Conversely, if a finitely additive measure μ on $\mathcal{F}$ is σ-additive on each $p_n^{-1}(\mathcal{B}_n)$, then $\mu_n=p_n\circ\mu$ forms a self-consistent sequence of measures. In other words, (5.6) gives a one-to-one correspondence between the self-consistent sequences $\{\mu_n\}$ and the finitely additive measures μ on $\mathcal{F}$ which are σ-additive on each $p_n^{-1}(\mathcal{B}_n)$. Therefore Problem A can be written in the following way.

Problem B. Let $\{\mathcal{B}'_n\}$ be an increasing family of σ-algebras. Suppose that μ is σ-additive on each $\mathcal{B}'_n$ but finitely additive on $\mathcal{F}=\bigcup_{n=1}^{\infty}\mathcal{B}'_n$. Can μ be extended to a σ-additive measure on $\mathcal{B}(\mathcal{F})$?

If μ is extendable, the extension is unique by Th. 3.4.

Problem B seems to be more general, but actually is not so. In the setting of Prob. B, if we denote the identity map with id, $\{(X,\mathcal{B}'_n),id\}$ becomes a projective sequence of measurable spaces, and their projective limit is isomorphic with $(X,\mathcal{B}(\mathcal{F}))$. Since p_n is also the identity map, Prob. A for this sequence is exactly the same with Prob. B. Thus, the apparently general Prob. B is merely an example of Prob. A, and two problems are equivalent.

From the next section on, we shall study some sufficient conditions for the extendability. But in order to show that our problem is not trivial, in this section we shall give examples in which the extension is not assured.

<u>Example 5.1.</u> Let X be an infinite set. There exists a decreasing sequence $\{A_n\}$ of non-empty subsets such that $\bigcap_{n=1}^{\infty} A_n=\emptyset$. Put $\mu(E)=1$ if $E\supset A_n$ for some n, $=0$ otherwise. Let $\mathcal{B}'_n$ be the σ-algebra generated by $\{A_1,A_2,\cdots,A_n\}$. (Actually $\mathcal{B}'_n$ consists of finite number of subsets). Then μ is σ-additive on each $\mathcal{B}'_n$, therefore finitely additive on $\mathcal{F} = \bigcup_{n=1}^{\infty}\mathcal{B}'_n$. But $\mu(A_n)=1$ and $\bigcap_{n=1}^{\infty} A_n=\emptyset$ imply that μ can not be extended to a measure on $\mathcal{B}(\mathcal{F})$.

<u>Example 5.2.</u> Let $(X,\mathcal{B},\mu)$ be a measure space, and $\{A_n\}$ be a decreasing sequence of thick subsets such that $\bigcap_{n=1}^{\infty} A_n=\emptyset$. Let $\mathcal{B}'_n$ be the smallest σ-algebra containing $\mathcal{B}$ and $A_1,A_2,\cdots,A_n$. μ can be regarded as a measure on $\mathcal{B}'_n$. (First take the trace on A_n, then imbed into X). Again $\mu(A_n)=1$ and $\bigcap_{n=1}^{\infty} A_n=\emptyset$ imply that μ can not be extended to a measure on $\mathcal{B}(\mathcal{B}\cup\{A_n\})$.

Let $X=[0,1)$, $\mathcal{B}$ be the Borel field on $[0,1)$, and μ be the usual Lebesgue measure. Let H be a set such that $\{H+r \mid r \text{ is a rational number}\}$ forms a countable partition of $[0,1)$ (in mod. 1). Then putting $A_n=\bigcap_{k=1}^{n}(H+r_k)^c$, we have a special example of the above one.

Example 5.3. In the next chapter, we shall discuss the extendability for real vector spaces. It turns out that the extendability is assured only under some restrictive conditions. In other words, without such conditions, our problem B is negative.

Example 5.4. Counter example for direct product.

Let $\{(X^{(k)},\mathcal{B}^{(k)})\}$ be a sequence of measurable spaces, and μ be a measure on the product measurable space $(X,\mathcal{B})$. Evidently, $p_n \circ \mu=\mu_n$ forms a self-consistent sequence on the partial product $(X_n,\mathcal{B}_n)$.

Suppose that for some $A^{(k)}\subset X^{(k)}$, $Y_n=\prod_{k=1}^{n} A^{(k)}$ is thick in μ_n, but $Y=\prod_{k=1}^{\infty} A^{(k)}$ is not thick in μ. Then, taking the trace on Y_n, μ_n can be regarded as a measure on Y_n, and $\{\mu_n\}$ becomes a self-consistent sequence on the partial product Y_n. If it were extendable to a measure on Y, the extended measure would be regarded as a measure on X by imbedding, but from the uniqueness of the extension, it must be equal with μ. Thus, Y should be thick in μ. But this contradicts with the assumption, so $\{\mu_n\}$ can not be extended to a measure on $(Y,\mathcal{B}\cap Y)$.

Consider the following special case. Suppose that $X^{(k)}=[0,1)$ and $\mathcal{B}^{(k)}$ is the Borel field on $[0,1)$ (which we denote with $\mathcal{B}_0$). Then we have $X=[0,1)^\infty$, and $\mathcal{B}=\mathcal{B}_0^\infty$. Since $([0,1),\mathcal{B}_0)$ is isomorphic with $(\Delta,\mathcal{B}\cap\Delta)$ where Δ is the diagonal in $[0,1)^\infty$, the Lebesgue measure ν can be regarded as a measure on Δ, hence as a measure μ on $(X,\mathcal{B})$ by imbedding. Put $A^{(k)}=(H+r_k)^c$, where H is the set given in Ex. 5.2. Then this gives a special example of the above one. ($\prod_{k=1}^{n} A^{(k)}$ is thick in $\dot{\mu}_n$ because $\bigcap_{k=1}^{n} A^{(k)}$ is thick in ν, while $\prod_{k=1}^{\infty} A^{(k)}\cap\Delta=\emptyset$ implies that $\prod_{k=1}^{\infty} A^{(k)}$ is not thick in μ, because $\Delta\in\mathcal{B}$ is assured by Def. 7.1 and Th. 7.1).

In this example, each $((H+r_k)^c,\mathcal{B}_0\cap(H+r_k)^c)$ is isomorphic with $(H^c,\mathcal{B}_0\cap H^c)$. Therefore this gives a counter example for a stationary product measurable space (the case in which all $(X^{(k)},\mathcal{B}^{(k)})$ are mutually isomorphic).

§6. Extension theorem for direct product

Definition 6.1. Let $(X,\mathcal{B},\mu)$ be a measure space. The outer measure μ^* and the inner measure μ_* are defined on 2^X as follows:

$$(6.1)\qquad \mu^*(E) = \inf\{\mu(B)\mid B\in\mathcal{B} \text{ and } B\supset E\},$$

$$(6.2)\qquad \mu_*(E) = \sup\{\mu(B)\mid B\in\mathcal{B} \text{ and } B\subset E\}.$$

Evidently we have $\mu^*=\mu_*=\mu$ on $\mathcal{B}$. μ^* is a Caratheodory's outer measure and $\mu_*(E)=\mu(X)-\mu^*(E^c)$. The function $d^*(E_1,E_2)=\mu^*(E_1\Delta E_2)$ is a metric on 2^X (after identifying E_1 and E_2 such that $\mu^*(E_1\Delta E_2)=0$), and we have $d^*(E_1,E_2)=d^*(E_1^c,E_2^c)$ as well as $d^*(\bigcup_{i=1}^{\infty} E_{1i},\bigcup_{i=1}^{\infty} E_{2i})\leq\sum_{i=1}^{\infty} d^*(E_{1i},E_{2i})$.

Definition 6.2. Let $(X,\mathcal{B},\mu)$ be a measure space, and τ be a topology on X. The compact inner measure μ_τ is defined on 2^X as follows:

$$(6.3) \qquad \mu_\tau(E) = \sup\{\mu_*(K) \mid K \text{ is compact and } K\subset E\}.$$

μ is said to be compact regular in τ, if $\mu=\mu_\tau$ on $\mathcal{B}$ (or equivalently $\mu_*=\mu_\tau$ on 2^X).

μ is compact regular if and only if

$$(6.4) \qquad \forall B\in\mathcal{B}, \quad \inf\{d^*(B,K) \mid K \text{ is compact and } K\subset B\} = 0.$$

Theorem 6.1. Let $\{(X^{(k)},\mathcal{B}^{(k)})\}$ be a sequence of measurable spaces, and $\{\mu_n\}$ be a self-consistent sequence of measures on the partial products. For some topology $\tau^{(k)}$ on $X^{(k)}$, suppose that μ_n is compact regular in the product topology $\tau_n=\tau^{(1)}\times\tau^{(2)}\times\cdots\times\tau^{(n)}$. Then, $\{\mu_n\}$ can be extended to a σ-additive measure on the infinite product measurable space.

Remark. Even if τ_n is not a product topology, the same result holds. (cf. Th. 7.2).

Proof. We shall apply Hopf's theorem. We want to prove that for every decreasing sequence $\{F_n\}$ in $\mathcal{F}=\bigcup_{n=1}^{\infty} p_n^{-1}(\mathcal{B}_n)$, we have

$$(6.5) \qquad \lim_{n\to\infty} \mu(F_n)=\alpha>0 \Longrightarrow \bigcap_{n=1}^{\infty} F_n \neq \emptyset$$

$F_n\in\mathcal{F}$ means that $\exists m_n$, $F_n\in p_{m_n}^{-1}(\mathcal{B}_{m_n})$, but since $\{p_n^{-1}(\mathcal{B}_n)\}$ is increasing, we can assume that $\{m_n\}$ is increasing. Put F'_m $=X$ if $m<m_1$, and $=F_n$ if $m_n\leq m<m_{n+1}$. Then we have $F'_m \in p_m^{-1}(\mathcal{B}_m)$, $\lim_{m\to\infty}\mu(F'_m)=\alpha$ and $\bigcap_{m=1}^{\infty}F'_m=\bigcap_{n=1}^{\infty}F_n$. Therefore, it is sufficient to prove (6.5) under the assumption $\forall n$, $F_n\in p_n^{-1}(\mathcal{B}_n)$.

Assume $F_n \in p_n^{-1}(\mathcal{B}_n)$, then $\exists B_n \in \mathcal{B}_n$, $F_n = p_n^{-1}(B_n)$. Since μ_n is compact regular, there exists a compact set K_n contained in B_n such that $d_n^*(B_n, K_n) < \frac{\alpha}{2^n}$. Then we have

$$(6.6) \qquad d_n^*(B_n, \bigcap_{k=1}^{n} p_{kn}^{-1}(K_k)) \leqq \sum_{k=1}^{n} d_n^*(p_{kn}^{-1}(B_k), p_{kn}^{-1}(K_k))$$

$$\leqq \sum_{k=1}^{n} d_k^*(B_k, K_k) < \sum_{k=1}^{n} \frac{\alpha}{2^k} < \alpha.$$

Thus we get $\bigcap_{k=1}^{n} p_{kn}^{-1}(K_k) \neq \emptyset$, hence $\bigcap_{k=1}^{n} p_k^{-1}(K_k) \neq \emptyset$. Evidently we have $\bigcap_{n=1}^{\infty} F_n = \bigcap_{n=1}^{\infty} p_n^{-1}(B_n) \supset \bigcap_{n=1}^{\infty} p_n^{-1}(K_n)$, so that the proof will be completed using the following lemma.

<u>Lemma 6.1.</u> Let $\{(X^{(k)}, \tau^{(k)})\}$ be a sequence of topological spaces, and K_n be a compact set in the partial product topological space (X_n, τ_n). Then the finite intersection property of $\{p_n^{-1}(K_n)\}$ implies the complete intersection property.

<u>Proof.</u> Put $K^{(n)} = p_n^{(n)}(K_n)$, then $K^{(n)}$ is compact in $\tau^{(n)}$ and $K_n \subset p_n^{(n)-1}(K^{(n)})$, thus we have $\bigcap_{k=1}^{n} p_{kn}^{-1}(K_k) \subset \bigcap_{k=1}^{n} p_n^{(k)-1}(K^{(k)}) = \prod_{k=1}^{n} K^{(k)}$.

From the finite intersection property of $\{p_n^{-1}(K_n)\}$, we have $\bigcap_{k=1}^{n} p_{kn}^{-1}(K_k) \neq \emptyset$. On the other hand, the projection p_n maps $\prod_{k=1}^{\infty} K^{(k)}$ onto $\prod_{k=1}^{n} K^{(k)}$, so that $p_n^{-1}(\bigcap_{k=1}^{n} p_{kn}^{-1}(K_k)) \cap \prod_{k=1}^{\infty} K^{(k)} \neq \emptyset$, namely we have $\bigcap_{k=1}^{n} p_k^{-1}(K_k) \cap K \neq \emptyset$ where $K = \prod_{k=1}^{\infty} K^{(k)}$.

By Tychonov's theorem, K is compact. Since every $p_k^{-1}(K_k)$ is a closed set, from the definition of a compact set we have $\bigcap_{k=1}^{\infty} p_k^{-1}(K_k) \cap K \neq \emptyset$. (q.e.d.)

We shall denote the extended measure on $\mathfrak{B}(\mathfrak{F})$ with the same symbol μ.

Theorem 6.2. Assume the same situation with Th. 6.1. If Y_n is thick in μ_n, then in either of the following two cases, $\bigcap_{n=1}^{\infty} p_n^{-1}(Y_n)$ is thick in μ.

(1) $\forall n,\quad Y_n \in \mathcal{B}_n$

(2) Y_n is closed in τ_n, and $\{p_n^{-1}(Y_n)\}$ is decreasing.

Proof. The case (1) is simple. Since Y_n is thick and $\mathcal{B}_n$-measurable, we have $\mu_n(Y_n^c)=0$, namely $\mu(p_n^{-1}(Y_n)^c)=0$. Hence we have $\mu(\bigcup_{n=1}^{\infty} p_n^{-1}(Y_n)^c)=0$, thus $\bigcap_{n=1}^{\infty} p_n^{-1}(Y_n)$ is thick in μ.

The proof for the case (2) is as follows. Assuming $B\in\mathcal{B}$, $\mu(B)=\alpha>0$, we shall prove $B\cap\bigcap_{n=1}^{\infty} p_n^{-1}(Y_n)\neq\emptyset$.

First consider the case of $B=\bigcap_{n=1}^{\infty} p_n^{-1}(B_n)$, $B_n\in\mathcal{B}_n$. Since μ_n is compact regular, there exists a compact set K_n contained in B_n such that $d_n^*(B_n,K_n)<\frac{\alpha}{2^n}$. Then similarly as in (6.6), we have $\mu_{n*}(\bigcap_{k=1}^{n} p_{kn}^{-1}(K_k))>0$. Since Y_n is thick in μ_n, this implies $Y_n\cap\bigcap_{k=1}^{n} p_{kn}^{-1}(K_k)\neq\emptyset$, so that $p_n^{-1}(Y_n)\cap\bigcap_{k=1}^{n} p_k^{-1}(K_k)\neq\emptyset$. But $\{p_n^{-1}(Y_n)\}$ is decreasing, so that $\bigcap_{k=1}^{n} p_k^{-1}(Y_k\cap K_k)\neq\emptyset$. Since $Y_n\cap K_n$ is a compact set in τ_n, from Lemma 6.1 we see that $\bigcap_{n=1}^{\infty} p_n^{-1}(Y_n\cap K_n)\neq\emptyset$, therefore remarking $\bigcap_{n=1}^{\infty} p_n^{-1}(K_n)\subset \bigcap_{n=1}^{\infty} p_n^{-1}(B_n)=B$, we get $\bigcap_{n=1}^{\infty} p_n^{-1}(Y_n)\cap B\neq\emptyset$.

The proof for general $B\in\mathcal{B}$ can be reduced to the above one using the following lemma.

Lemma 6.2. Let $\mathcal{F}$ be a finite algebra, and $\mathcal{F}_\delta$ be the family of all countable intersections of sets in $\mathfrak{F}$: $\mathfrak{F}_\delta=\{\bigcap_{n=1}^{\infty} F_n \mid F_n\in\mathfrak{F}\}$. Let μ be a σ-additive measure on $\mathcal{B}(\mathfrak{F})$, then for every $B\in \mathcal{B}(\mathfrak{F})$ we have

(6.7) $\qquad \mu(B) = \sup\{\mu(E) \mid E\in\mathfrak{F}_\delta \text{ and } E\subset B\}.$

Proof. Let $\mathfrak{A}$ be the family of all $\mathcal{B}(\mathfrak{F})$-measurable sets which satisfy (6.7). Evidently $\mathfrak{A}$ contains $\mathfrak{F}$. Since $\mathfrak{F}_\delta$ is closed under countable intersection and finite union, $\mathfrak{A}$ is closed under countable intersection and countable union. So that $\mathfrak{A}\cap\mathfrak{A}^c$ is a σ-algebra (where $\mathfrak{A}^c=\{A^c \mid A\in\mathfrak{A}\}$). Hence $\mathfrak{A}\cap\mathfrak{A}^c\supset \mathcal{B}(\mathfrak{F})$, thus combining with $\mathfrak{A}\subset\mathcal{B}(\mathfrak{F})$, we see $\mathfrak{A}=\mathcal{B}(\mathfrak{F})$.

(q.e.d.)

§7. Extension theorem for projective limit

Let X be a set, and E be a subset. We shall denote the indicator function of E with c_E, namely $c_E(x)=1$ if $x\in E$, $=0$ if $x\notin E$. For $x\neq x'$, put

(7.1) $\qquad \mathcal{B}_{x,x'} = \{E\in 2^X \mid c_E(x) = c_E(x')\},$

then $\mathcal{B}_{x,x'}$ becomes a σ-algebra.

Definition 7.1. Let X be a set, and $\mathfrak{A}$ be a family of subsets of X. $\mathfrak{A}$ is said to separate X (or X is separated by $\mathfrak{A}$) if $x\neq x'$ implies $\mathfrak{A}\not\subset\mathcal{B}_{x,x'}$, namely $\exists A\in\mathfrak{A}$, $x\in A$, $x'\notin A$.

A measurable space $(X,\mathcal{B})$ is said to be σ-separated, if X is separated by some countable family $\mathfrak{A}\subset\mathcal{B}$.

Definition 7.2. A σ-algebra $\mathcal{B}$ is said to be σ-generated, if $\mathcal{B}$ is generated by some countable family $\mathfrak{A}$, namely $\mathcal{B}=\mathcal{B}(\mathfrak{A})$.

In general, if X is separated by $\mathcal{B}(\mathfrak{A})$, then X is already separated by $\mathfrak{A}$. Especially, if $\mathcal{B}$ is σ-generated and if every one-point set belongs to $\mathcal{B}$, then $(X,\mathcal{B})$ is σ-separated.

<u>Theorem 7.1.</u> Let φ be a measurable surjective map from $(X, \mathcal{B})$ to $(X', \mathcal{B}')$. The graph $G(\varphi)=\{(x, \varphi(x)) \mid x \in X\}$ is $\mathcal{B} \times \mathcal{B}'$-measurable if and only if $(X', \mathcal{B}')$ is σ-separated.

<u>Corollary</u>. The diagonal in X^2 is $\mathcal{B}\times\mathcal{B}$-measurable if and only if $(X,\mathcal{B})$ is σ-separated. (This result can be easily generalized to the diagonal in finite or countable direct product).

<u>Proof</u> (of Th.). Suppose that X' is separated by $\{B_n'\} \subset \mathcal{B}'$, and denote the indicator function of B_n' with c_n. It is a measurable function (namely, a measurable map from X' to $\mathbb{R}$ with respect to $\mathcal{B}'$ and the usual Borel field on $\mathbb{R}$). Since the projections are measurable, the following f is a measurable function on $(X\times X', \mathcal{B}\times\mathcal{B}')$.

$$(7.2) \qquad f(x,x') = \sum_{n=1}^{\infty} | c_n(\varphi(x)) - c_n(x')|.$$

Therefore $f^{-1}(0) \in \mathcal{B} \times \mathcal{B}'$. But $(x,x') \in f^{-1}(0)$ means that $\forall n$, $c_n(\varphi(x))=c_n(x')$, thus we get $\varphi(x)=x'$. Therefore we have $f^{-1}(0)=G(\varphi)$, so $G(\varphi) \in \mathcal{B} \times \mathcal{B}'$.

Before the proof of the converse, we shall remark two facts. As easily checked (or as well known), $E \in \mathcal{B}\times\mathcal{B}'$ implies that its section $E(x)=\{x' \in X' \mid (x,x') \in E\}$ belongs to $\mathcal{B}'$. Secondly, $E \in \mathcal{B}\times\mathcal{B}'$ implies that for some σ-generated σ-algebras $\mathcal{B}_1 \subset \mathcal{B}$ and $\mathcal{B}_1' \subset \mathcal{B}'$, E already belongs to $\mathcal{B}_1 \times \mathcal{B}_1'$ as proved in the next lemma.

Suppose that $G(\varphi) \in \mathcal{B}\times\mathcal{B}'$. Then for some $\mathcal{B}_1$ and $\mathcal{B}_1'$, we have $G(\varphi) \in \mathcal{B}_1 \times \mathcal{B}_1'$, therefore $G(\varphi)(x) \in \mathcal{B}_1'$. However, $G(\varphi)(x)$ is a one-point set $\{\varphi(x)\}$, so that every one-point set of X' belongs to $\mathcal{B}_1'$. Thus $(X', \mathcal{B}_1')$ is σ-separated, hence $(X', \mathcal{B}')$ is so.

Lemma 7.1. Let $\mathcal{O}\!\mathcal{l}$ be a family of subsets of X, then we have

$$(7.3) \qquad \mathcal{B}(\mathfrak{A}) = \bigcup_{\mathfrak{A}'} \mathcal{B}(\mathfrak{A}'),$$

where $\mathfrak{A}'$ runs over all countable families contained in $\mathfrak{A}$.

Proof. Denote the right side of (7.3) with $\mathcal{B}'$. Evidently we have $\mathfrak{A} \subset \mathcal{B}' \subset \mathcal{B}(\mathfrak{A})$. We shall show that $\mathcal{B}'$ is a σ-algebra. $\mathcal{B}'$ is closed under complément, because each $\mathcal{B}(\mathfrak{A}')$ is so. Suppose $B_k \in \mathcal{B}(\mathfrak{A}_k')$, where $\mathfrak{A}_k'$ is a countable family $\subset \mathfrak{A}$, then $\bigcup_{k=1}^{\infty} \mathfrak{A}_k' = \mathfrak{A}'$ is also a countable family $\subset \mathfrak{A}$ and $\forall k$, $B_k \in \mathcal{B}(\mathfrak{A}')$, therefore $\bigcup_{k=1}^{\infty} B_k \in \mathcal{B}(\mathfrak{A}') \subset \mathcal{B}'$. (q.e.d.)

Theorem 7.2. Let $\{(X_n, \mathcal{B}_n), p_{nm}\}$ be a projective sequence of measurable spaces, and let $\{\mu_n\}$ be a self-consistent sequence of measures. Suppose that μ_n is compact regular in some topology τ_n on X_n, then $\{\mu_n\}$ is extendable to a σ-additive measure on the projective limit measurable space, in either of the following two cases:

(1) $(X_n, \mathcal{B}_n)$ is σ-separated (with no condition on τ_n).

(2) p_{nm} is continuous with respect to τ_m and τ_n (with no condition on $\mathcal{B}_n$).

Proof. Let $(X, \mathcal{B})$ be the direct product of $\{(X_n, \mathcal{B}_n)\}$, and $(X^m, \mathcal{B}^m)$ be the partial product. Let π_n, π^m, π^{nm} and π_n^m be the projections from X to X_n, from X to X^m, from X^m to $X^n (n<m)$, and from X^m to $X_n (n \leq m)$. (Note that the superscripts and subscripts are reversed comparing with in §4).

Let f_m be a map from X_m to X^m defined by

(7.4) $$f_m(x_m) = (p_{1m}(x_m), p_{2m}(x_m), \cdots, x_m)$$

f_m is measurable with respect to $\mathcal{B}_m$ and $\mathcal{B}^m$. Denote the image measure $f_m \circ \mu_m$ with ν^m. Then for $\ell > m$, we have $\pi^{m\ell} \circ \nu^\ell = \pi^{m\ell} \circ f_\ell \circ \mu_\ell = f_m \circ p_{m\ell} \circ \mu_\ell = f_m \circ \mu_m = \nu^m$, thus $\{\nu^m\}$ is self-consistent.

ν^m is compact regular in $\tau^m = \tau_1 \times \tau_2 \times \cdots \times \tau_m$ as proved below. Let $\mathfrak{A}^m$ be the family of all $B^m \in \mathcal{B}^m$ which satisfy $\nu^m(B^m) = \sup\{\nu^m_*(K^m) \mid K^m$ is compact in τ^m and $K^m \subset B^m\}$. Then $\mathfrak{A}^m$ is closed under countable intersection and countable union. On the other hand for $B_k \in \mathcal{B}_k$, we have $\prod_{k=1}^m B_k \in \mathfrak{A}^m$, because $d^{m*}(\prod_{k=1}^m B_k, \prod_{k=1}^m K_k) \leqq \sum_{k=1}^m d^{m*}((\pi_k^m)^{-1}(B_k), (\pi_k^m)^{-1}(K_k)) \leqq \sum_{k=1}^m d^{k*}((\pi_k^k)^{-1}(B_k), (\pi_k^k)^{-1}(K_k)) = \sum_{k=1}^m d_k^*(B_k, K_k)$. Thus similarly as in the proof of Lemma 6.2, we see that $\mathfrak{A}^m = \mathcal{B}^m$.

By Th. 6.1, $\{\nu^m\}$ can be extended to a σ-additive measure ν on $(X, \mathcal{B})$: $\pi^m \circ \nu = \nu^m$. Put $Y^m = f_m(X_m)$, then evidently Y^m is thick in ν^m. Since $Y^m = \bigcap_{k=1}^m \{(x_1, x_2, \cdots, x_m) \mid x_k = p_{km}(x_m)\}$, if $(X_k, \mathcal{B}_k)$ is σ-separated, then Y^m is $\mathcal{B}^m$-measurable by Th. 7.1, and if p_{km} is continuous, then Y^m is closed in τ^m. Evidently $\{(\pi^m)^{-1}(Y^m)\}$ is decreasing and we have $\bigcap_{m=1}^\infty (\pi^m)^{-1}(Y^m) = X_\infty$. Therefore, in either of two cases (1) or (2), X_∞ becomes a thick set in ν by Th. 6.2.

Denote the trace of ν on X_∞ with μ, and the injection $X_\infty \to X$ with i. Then we have $p_n \circ \mu = \pi_n \circ i \circ \mu = \pi_n \circ \nu = \pi_n^n \circ \pi^n \circ \nu = \pi_n^n \circ \nu^n = \pi_n^n \circ f_n \circ \mu_n = \mu_n$, so μ is a σ-additive extension of $\{\mu_n\}$ on $(X_\infty, \mathcal{B}_\infty)$. (q.e.d.)

Remark 1. A more careful discussion shows that the conditions (1) and (2) in Th. 7.2 can be replaced by: there exists a thick $\mathcal{B}_n$-measurable set $Y_n \subset X_n$ such that (1)' $(Y_n, \mathcal{B}_n \cap Y_n)$ is σ-separated, or (2)' p_{mn} is continuous on Y_n.

Remark 2. Even if $\{\nu^m\}$ can be extended to a measure ν on $(X, \mathcal{B})$, X_∞ is not always thick in ν. A counter example is given by Example 5.1. In this case, ν exists but we have $\nu(\prod_{k=1}^{\infty} A_k)=1$ so that the diagonal is not thick.

§8. Non-countable direct product and projective limit

Definition 8.1. Let $\{(X^{(\lambda)}, \mathcal{B}^{(\lambda)})\}_{\lambda \in \Lambda}$ be a family of measurable spaces. The product measurable space $(X, \mathcal{B})$ is defined as follows: $X = \prod_{\lambda \in \Lambda} X^{(\lambda)}$ and $\mathcal{B}$ is the smallest σ-algebra in which all projections $p^{(\lambda)}: X \to X^{(\lambda)}$ are measurable, namely

$$(8.1) \qquad \mathcal{B} = \mathcal{B}(\bigcup_{\lambda \in \Lambda} p^{(\lambda)-1}(\mathcal{B}^{(\lambda)})).$$

Let L be a subset of Λ. We shall define the partial product $(X_L, \mathcal{B}_L)$ in a similar way: $X_L = \prod_{\lambda \in L} X^{(\lambda)}$, $\mathcal{B}_L = \mathcal{B}(\bigcup_{\lambda \in L} p_L^{(\lambda)-1}(\mathcal{B}^{(\lambda)}))$, where $p_L^{(\lambda)}$ is the projection from X_L to $X^{(\lambda)} (\lambda \in L)$. Let p_L and $p_{LL'}$ be the projections from X to X_L, and from $X_{L'}$ to $X_L (L \subset L')$.

Let $\mathcal{L}$ be the family of all finite subsets of Λ. Then we have

$$(8.2) \qquad \mathcal{B} = \mathcal{B}(\mathcal{F}), \quad \mathcal{F} = \bigcup_{L \in \mathcal{L}} p_L^{-1}(\mathcal{B}_L),$$

$\mathcal{F}$ being a finite algebra.

Let μ be a measure on $(X, \mathcal{B})$. Then putting $\mu_L = p_L \circ \mu$, we get a family $\{\mu_L\}_{L\in\mathcal{L}}$ which satisfies the self-consistency condition:

(8.3) $\mu_L = p_{LL'}\circ\mu_{L'}$ for $L\subset L'$.

The converse is our problem. Suppose that a self-consistent family $\{\mu_L\}_{L\in\mathcal{L}}$ is given. Our problem can be reduced to the case of countable direct product as explained below. For a sequence $\{\lambda_k\}\subset\Lambda$, put $L_n=\{\lambda_1,\lambda_2,\cdots,\lambda_n\}$, then $\{\mu_{L_n}\}$ is a self-consistent sequence in the sense of §5.

Theorem 8.1. Let $\{(X^{(\lambda)}, \mathcal{B}^{(\lambda)})\}_{\lambda\in\Lambda}$ be a family of measurable spaces and let $\{\mu_L\}_{L\in\mathcal{L}}$ be a self-consistent family of measures on finite partial products. $\{\mu_L\}$ is extendable to a measure on $(X,\mathcal{B})$, if and only if $\{\mu_{L_n}\}$ is extendable for every sequence $\{\lambda_k\}$.

Proof. Let μ be an extension of $\{\mu_L\}$, namely $\mu_L=p_L\circ\mu$. Then, for a given sequence $\{\lambda_k\}$, putting $M=\{\lambda_1,\lambda_2,\cdots\}$ and denoting $p_M\circ\mu$ with μ_M, we have $p_{L_nM}\circ\mu_M=p_{L_nM}\circ p_M\circ\mu=p_{L_n}\circ\mu=\mu_{L_n}$. Thus μ_M is an extension of $\{\mu_{L_n}\}$.

Conversely, suppose that for every sequence $\{\lambda_k\},\{\mu_{L_n}\}$ can be extended to a measure μ_M on $(X_M, \mathcal{B}_M)$: $p_{L_nM}\circ\mu_M=\mu_{L_n}$. If L is a finite subset of M, then we have $L\subset L_n$ for some n, so that $p_{LM}\circ\mu_M=p_{LL_n}\circ p_{L_nM}\circ\mu_M=p_{LL_n}\circ\mu_{L_n}=\mu_L$. Hence from the uniqueness of the extension, we see that μ_M does not depend on the choice of the order of $\{\lambda_k\}$, but depends only on the set $M=\{\lambda_k\}$.

Let $\mathfrak{M}$ be the family of all countable subsets of Λ. $M\subset M'$ implies that for every $L\subset M$, we have $\mu_L=p_{LM'}\circ\mu_{M'}=p_{LM}\circ$

$p_{MM'} \circ \mu_{M'}$, therefore from the uniqueness of the extension we get $\mu_M = p_{MM'} \circ \mu_{M'}$. Thus $\{\mu_M\}_{M \in \mathfrak{M}}$ is again a self-consistent family.

From Lemma 7.1, we have $\mathcal{B} = \bigcup_{M \in \mathfrak{M}} p_M^{-1}(\mathcal{B}_M)$. We shall define μ by:

$$(8.4) \qquad \mu(p_M^{-1}(B_M)) = \mu_M(B_M), \qquad \forall B_M \in \mathcal{B}_M.$$

This defines μ consistently, because $p_M^{-1}(B_M) = p_{M'}^{-1}(B_{M'})$ implies $p_{M''}^{-1}(p_{MM''}^{-1}(B_M)) = p_{M''}^{-1}(p_{M'M''}^{-1}(B_{M'}))$ for $M'' = M \cup M'$, hence $p_{MM''}^{-1}(B_M) = p_{M'M''}^{-1}(B_{M'})$, hence $\mu_M(B_M) = \mu_{M'}(B_{M'})$. μ is σ-additive on $\mathcal{B}$, because countable number of sets in $\mathcal{B}$ always belong to some common $p_M^{-1}(\mathcal{B}_M)$.

Finally μ is an extension of $\{\mu_L\}_{L \in \mathcal{L}}$, because for a given $L \in \mathcal{L}$, taking M which contains L, we have $p_L \circ \mu = p_{LM} \circ p_M \circ \mu = p_{LM} \circ \mu_M = \mu_L$. (Actually the extension is unique).

(q.e.d.)

Similar discussions can be applied to projective limit.

Definition 8.2. A semi-ordered set $\mathcal{L}$ is called a directed set, if

$$(8.5) \qquad \forall L_1, L_2 \in \mathcal{L}, \quad \exists L_3 \in \mathcal{L}, \quad L_1 \leq L_3 \text{ and } L_2 \leq L_3.$$

A maximal element of a directed set is necessarily the maximum element.

Definition 8.3. Let $\{(X_L, \mathcal{B}_L)\}_{L \in \mathcal{L}}$ be a family of measurable spaces parametrized by elements of a directed set $\mathcal{L}$. Suppose that for $L < L'$, a measurable surjective map $p_{LL'}$ is given with the condition $p_{LL''} = p_{LL'} \circ p_{L'L''}$ for $L < L' < L''$. Such a family $\{(X_L, \mathcal{B}_L), p_{LL'}\}$ is called a projective family of measurable spaces. Its projective limit measurable space $(X_\infty, \mathcal{B}_\infty)$ is defined as follows: Let $(X, \mathcal{B})$ be the direct product of $\{(X_L, \mathcal{B}_L)\}$. X_∞ is the following subset of X, and $\mathcal{B}_\infty = \mathcal{B} \cap X_\infty$.

$$(8.6) \qquad X_\infty = \{(x_L) \mid L < L' \Longrightarrow x_L = p_{LL'}(x_{L'})\}.$$

We shall denote the projection from X to X_L with π_L, and its restriction on X_∞ with p_L. Evidently we have (8.2) replacing $\mathcal{B}$ by $\mathcal{B}_\infty$.

Let μ be a measure on $(X_\infty, \mathcal{B}_\infty)$, and put $\mu_L = p_L \circ \mu$. Then $\{\mu_L\}$ satisfies the self-consistency condition (8.3) replacing $L \subset L'$ by $L < L'$.

The converse is our problem. Suppose that a self-consistent family $\{\mu_L\}$ is given. If $\mathcal{L}$ has the maximum element L_0, then $(X_{L_0}, \mathcal{B}_{L_0})$ is isomorphic with $(X_\infty, \mathcal{B}_\infty)$, and the measure μ_{L_0} is an extension of $\{\mu_L\}$.

Hereafter we shall assume that $\mathcal{L}$ has no maximal element. Then, for an increasing sequence $L_1 < L_2 < \cdots < L_n < \cdots$, $\{(X_{L_n}, \mathcal{B}_{L_n}), p_{L_n L_m}\}$ is a projective sequence in the sense of §4. We shall denote its projective limit measurable space with $(X_M, \mathcal{B}_M)$, where $M = \{L_n\}$ is a subset of $\mathcal{L}$. Let π^M be the projection from $X = \prod_{L \in \mathcal{L}} X_L$ to $\prod_{L \in M} X_L$, and let p_M be the restriction of π^M on X_∞. Evidently p_M maps X_∞ to X_M, but not necessarily surjective.

<u>Theorem 8.2.</u> Let $\{(X_L, \mathcal{B}_L), p_{LL'}\}_{L,L' \in \mathcal{L}}$ be a projective family of measurable spaces, and let $\{\mu_L\}$ be a self-consistent

family. $\{\mu_L\}$ is extendable to a measure on $(X_\infty, \mathcal{B}_\infty)$, if and only if for every increasing sequence $M=\{L_n\}$, $\{\mu_{L_n}\}$ is extendable to a measure μ_M and $p_M(X_\infty)$ is thick in μ_M.

Proof. Suppose that μ is an extension of $\{\mu_L\}$, namely assume $p_L \circ \mu = \mu_L$, then denoting $p_M \circ \mu$ with μ_M, we have $p_{L_n M} \circ \mu_M = p_{L_n M} \circ p_M \circ \mu = p_{L_n} \circ \mu = \mu_{L_n}$. This means that μ_M is an extension of $\{\mu_{L_n}\}$. ($p_{L_n M}$ is the restriction of $\pi^M_{L_n}$ on X_M, where $\pi^M_{L_n}$ is the projection from $\prod_{n=1}^{\infty} X_{L_n} = \prod_{L \in M} X_L$ to X_{L_n}). Evidently $p_M(X_\infty)$ is thick in μ_M.

Conversely, denote the family of all increasing sequences in $\mathcal{L}$ with $\mathfrak{M}$, and suppose that an extension μ_M exists for $\forall M \in \mathfrak{M}$.

First, we shall define a semi-order on $\mathfrak{M}$. Let $M=\{L_n\}$ and $M'=\{L'_n\}$, then $M \leqq M'$ means that $\forall n$, $\exists m$, $L_n \leqq L'_m$. If $M \leqq M'$, a map $p_{MM'}$ is defined from $X_{M'}$ to X_M as follows:

$$(8.7) \qquad p_{L_n M} \circ p_{MM'} = p_{L_n L'_m} \circ p_{L'_m M'}$$

This defines $p_{MM'}$ consistently, because $L_n < L'_m < L'_\ell$ implies $p_{L_n L'_m} \circ p_{L'_m M'} = p_{L_n L'_\ell} \circ p_{L'_\ell M'}$. (8.7) also implies $p_M = p_{MM'} \circ p_{M'}$.

From $L \in M \leqq M'$, we can derive for some $L' \in M'$, $p_{LM} \circ p_{MM'} \circ \mu_{M'} = p_{LL'} \circ p_{L'M'} \circ \mu_{M'} = p_{LL'} \circ \mu_{L'} = \mu_L$. Therefore from the uniqueness of the extension, we have $\mu_M = p_{MM'} \circ \mu_{M'}$. Thus $\{\mu_M\}_{M \in \mathfrak{M}}$ is again a self-consistent family.

If $M_k \in \mathfrak{M}$, $k=1,2,\cdots$ are given, we can find some element $M \in \mathfrak{M}$ which is larger than all M_k, as shown below. Let $M_k = \{L_{kn}\}$, then it is sufficient to take $M=\{L_n\}$ where $L_n > L_{1n}$,

$L_{2n}\cdots, L_{nn}, L_{n-1}$. From this we see that the family $\bigcup_{M\in\mathfrak{M}} p_M^{-1}(\mathcal{B}_M)$ is a σ-algebra, because countable number of sets in this family always belong to some common $p_M^{-1}(\mathcal{B}_M)$. So we get $\mathcal{B} = \bigcup_{M\in\mathfrak{M}} p_M^{-1}(\mathcal{B}_M)$.

We shall define a measure μ on $\mathcal{B}$ by

$$(8.8) \qquad \mu(p_M^{-1}(B_M)) = \mu_M(B_M) \qquad \forall B_M \in \mathcal{B}_M$$

This defines μ consistently, because $p_M^{-1}(B_M)=p_{M'}^{-1}(B_{M'})$ implies $p_{M''}^{-1}(p_{MM''}^{-1}(B_M))=p_{M''}^{-1}(p_{M'M''}^{-1}(B_{M'}))$ for $M''\geq M, M'$, and since $p_{M''}(X_\infty)$ is thick in $\mu_{M''}$, this implies $\mu_M(B_M)=\mu_{M'}(B_{M'})$. μ is σ-additive on $\mathcal{B}$, because countable number of sets in $\mathcal{B}$ always belong to some common $p_M^{-1}(\mathcal{B}_M)$.

Finally μ is an extension of $\{\mu_L\}$, because for a given L, taking $M\in\mathfrak{M}$ which contains L, we have $p_L\circ\mu=p_{LM}\circ p_M\circ\mu=p_{LM}\circ\mu_M=\mu_L$. (Actually the extension is unique). (q.e.d.)

§9. Compact regular measurable space

Definition 9.1. Let $(X,\mathcal{B})$ be a measurable space, and τ be a topology on X. $(X,\mathcal{B})$ is said to be compact regular in τ, if every measure on $\mathcal{B}$ is compact regular in τ.

Definition 9.2. a) Let (X,τ) be a topological space, and φ be a map from X to X'. The image topology $\varphi(\tau)$ is defined on X' as follows: O' is open in $\varphi(\tau)$ if and only if $\varphi^{-1}(O')$ is open in τ. ($\varphi(\tau)$ is Hausdorff if τ is Hausdorff and φ is injective).

b) Let X be a set, $\{Y_\lambda\}$ be a partition of X, and τ_λ be a topology on Y_λ. The sum topology τ is defined on X as follows: O is open in τ if and only if $O\cap Y_\lambda$ is open in

τ_λ for every λ.

Theorem 9.1. a) Suppose that $(X,\mathcal{B})$ is compact regular in τ, and that $(X,\mathcal{B})$ is isomorphic with $(X',\mathcal{B}')$ by a measurable isomorphism φ. Then $(X',\mathcal{B}')$ is compact regular in $\varphi(\tau)$.

b) Suppose that $(X,\mathcal{B})$ is compact regular in τ, and that $Y\in\mathcal{B}$ or Y is closed in τ, then $(Y,\mathcal{B}\cap Y)$ is compact regular in $\tau\cap Y$.

c) Let $\{Y_k\}$ be a countable or finite partition of X. Suppose that $Y_k\in\mathcal{B}$ and that $(Y_k,\mathcal{B}\cap Y_k)$ is compact regular in τ_k, then $(X,\mathcal{B})$ is compact regular in the sum topology τ.

d) Let $\{(X^{(k)},\mathcal{B}^{(k)})\}$ be a finite or countable family of measurable spaces. If each $(X^{(k)},\mathcal{B}^{(k)})$ is compact regular in $\tau^{(k)}$, then the product measurable space $(X,\mathcal{B})$ is compact regular in the product topology τ.

Proof. a) Let μ be a measure on $(X',\mathcal{B}')$, and consider $\nu=\varphi^{-1}\circ\mu$. Then $d^*_\mu(K',B')=d^*_\nu(\varphi^{-1}(K'),\varphi^{-1}(B'))$ assures our theorem, because K' is compact in $\varphi(\tau)$ if and only if $\varphi^{-1}(K')$ is compact in τ.

b) Let μ be a measure on $(Y,\mathcal{B}\cap Y)$, and denote the imbedding of μ in $(X,\mathcal{B})$ with ν. Then $d^*_\mu(K\cap Y,B\cap Y)\leqq d^*_\nu(K,B)$ assures our theorem, because: if Y is closed, then the compactness of K implies that of $K\cap Y$, and if $Y\in\mathcal{B}$, then putting $B_1=B\cap Y$ and choosing $K\subset B_1$, we have $d^*_\mu(K,B\cap Y)\leqq d^*_\nu(K,B_1)$.

c) Let μ be a measure on $(X,\mathcal{B})$, and μ_k be the restriction of μ on $\mathcal{B}\cap Y_k$. Then taking $K_k\subset B\cap Y_k$, we have $d^*_\mu(\bigcup_{k=1}^{n}K_k, B)\leqq\sum_{k=1}^{n}d^*_{\mu_k}(K_k,B\cap Y_k)+d_\mu(B\cap(\bigcup_{k=1}^{n}Y_k),B)$ (In the case of finite partition, we do not need the second term). The second term tends to 0 as $n\to\infty$, and $\bigcup_{k=1}^{n} K_k$ is compact in the sum topology τ. So we get our theorem.

d) Let μ be a measure on $(X,\mathcal{B})$. Let $\mathcal{Q}$ be the family of all $\mathcal{B}$-measurable sets B which satisfy: $\inf\{d_\mu^*(K,B) \mid K$ is compact and $K\subset B\}=0$. Then $\mathcal{Q}$ is closed under countable intersection and countable union. So by a similar discussion with Lemma 6.2, if we prove $\prod_k B^{(k)} \in \mathcal{Q}$ for $B^{(k)} \in \mathcal{B}^{(k)}$, then we get $\mathcal{Q} = \mathcal{B}$. Denote the image measure $p^{(k)}\circ\mu$ with $\mu^{(k)}$, then we have $d_\mu^*(\prod_k K^{(k)}, \prod_K B^{(k)}) \leq \sum_k d_\mu^*(p^{(k)-1}(K^{(k)}), p^{(k)-1}(B^{(k)})) \leq \sum_k d_k^*(K^{(k)}, B^{(k)})$. ($d_k^*$ is the outer metric corresponding to $\mu^{(k)}$). Thus we get our theorem. (q.e.d.)

Combining the results of §§6, 7, 8, we see that:

<u>Theorem 9.2.</u> Let $\{(X^{(\lambda)}, \mathcal{B}^{(\lambda)})\}_{\lambda\in\Lambda}$ be a family of measurable spaces, and suppose that each $(X^{(\lambda)}, \mathcal{B}^{(\lambda)})$ is compact regular in some topology $\tau^{(\lambda)}$. Then, every self-consistent family $\{\mu_L\}$ on the partial products can be extended to a σ-additive measure on the product measurable space $(X, \mathcal{B})$.

<u>Theorem 9.3.</u> Let $\{(X_n, \mathcal{B}_n), p_{nm}\}$ be a projective sequence of measurable spaces, and suppose that each $(X_n, \mathcal{B}_n)$ is compact regular in some topology τ_n. Then, in either of the following two cases, every self-consistent sequence $\{\mu_n\}$ can be extended to a σ-additive measure on the projective limit measurable space $(X_\infty, \mathcal{B}_\infty)$.

(1) The case that each $(X_n, \mathcal{B}_n)$ is σ-separated

(2) The case that each p_{nm} is continuous with respect to τ_m and τ_n.

This result is valid even for a projective family, if we assume that the map p_M defined before Th. 8.2 is surjective.

§10. Borel field

<u>Definition 10.1.</u> Let (X,τ) be a topological space, and let $\mathcal{B}$ be the smallest σ-algebra which contains all open sets. $\mathcal{B}$ is called the Borel field, and a set belonging to $\mathcal{B}$ is called a Borel set. $(X,\mathcal{B})$ is called the Borel measurable space associated with τ, and a measure on $\mathcal{B}$ is called a Borel measure.

<u>Definition 10.2.</u> A topological space (X,τ) is said to be σ-compact, if X can be expressed as a countable union of compact sets. (X,τ) is said to be separable, if there exists a dense countable set.

<u>Definition 10.3.</u> Let (X,d) be a metric space, namely a topological space defined by a metric d. The distance between a point x and a set A is defined by:

$$(10.1) \qquad d(x,A) = \inf_{y\in A} d(x,y).$$

The diameter of a set A is defined by:

$$(10.2) \qquad \delta(A) = \sup_{\substack{x\in A \\ y\in A}} d(x,y).$$

A σ-compact metric space is always separable, because for every $\varepsilon>0$, there exists a countable set A which satisfies $d(x,A)<\varepsilon$ for every $x\in X$.

<u>Theorem 10.1.</u> Let (X,d) be a σ-compact or complete separable, metric space. Then, the Borel measurable space $(X,\mathcal{B})$ is

σ-separated and compact regular in this topology.

Corollary. For a family of σ-compact or complete separable, metric spaces, every self-consistent family of Borel measures can be extended to a σ-additive measure on the product or projective limit measurable space. (In the case of non-countable projective limit, we assume that the map p_M defined before Th. 8.2 is surjective).

This corollary includes the case of the product of real lines, a classical form of Kolmogorov's extension theorem.

Example. The real line $\mathbb{R}$ (as well as n-dimensional Euclidean space $\mathbb{R}^n$) is σ-compact and complete.

A separable Banach space is complete, but not σ-compact.

The space of all rational numbers Q is σ-compact, but not complete in the induced topology from $\mathbb{R}$.

Proof (of Th.). First, we shall prove that $(X, \mathcal{B})$ is σ-separated. Let $\{x_n\}$ be a dense countable set in X, and O_{nm} be the open ball with the center at x_n and with the radius $1/m$. Suppose $x \neq y$, hence $d(x,y)=\alpha>0$. Then, taking m such that $1/m<\alpha/2$, and choosing x_n such that $x \in O_{nm}$, we have evidently $y \notin O_{nm}$. Thus $\{O_{nm}\}$ separates X.

Next, we shall prove that $(X, \mathcal{B})$ is compact regular, assuming that (X,d) is σ-compact. Let $\mathcal{A}$ be the family of all Borel sets B which satisfy:

$$(10.3) \qquad \inf\{d_\mu(K,B) \mid K \text{ is compact and } K \subset B\} = 0.$$

$\mathcal{A}$ is closed under countable intersection and countable union, so that in order to prove $\mathcal{A} = \mathcal{B}$, it is sufficient to show that $\mathcal{A}$ contains all open sets and all closed sets.

Evidently, a compact set is a Borel set, and belongs to $\mathfrak{A}$. A closed set F belongs to $\mathfrak{A}$, because F can be expressed as a countable union of compact sets: assuming $X=\bigcup_{n=1}^{\infty} K_n$, we have $F=\bigcup_{n=1}^{\infty}(F\cap K_n)$. An open set O belongs to $\mathfrak{A}$, because O can be expressed as a countable union of closed sets: $O=\bigcup_{n=1}^{\infty} F_n$, where

$$F_n = \{x \in X \mid d(x,O^c) \geq 1/n\}. \tag{10.4}$$

Thus the proof has been completed.

The proof for the complete separable case is as follows: Let $\{x_n\}$ be a dense countable set in X. If necessary, replacing d with an equivalent bounded metric, we can assume $\delta(X)\leq 1$. (Two metrics d and $d/(1+d)$ are equivalent, namely they induce the same topology on X). Consider the following map Φ:

$$\Phi: X \ni x \longrightarrow (d(x,x_n)) \in [0,1]^{\infty}. \tag{10.5}$$

Since $d(x,x_n)$ is a continuous function of x for each n, the map Φ is continuous with respect to the product topology on $[0,1]^{\infty}$. Φ is injective, because $\Phi(x)=\Phi(y)$ implies that $d(x,x_{n_k})\to 0$ is equivalent with $d(y,x_{n_k})\to 0$, hence $x=y$. Φ is homeomorphic, because for a given $\varepsilon>0$, choosing x_n such that $d(x,x_n)<\varepsilon/3$, we see that $|d(x,x_n)-d(y,x_n)|<\varepsilon/3$ implies $d(x,y)<\varepsilon$.

By Tychonov's theorem, $[0,1]^{\infty}$ is compact in the product topology. So, using the following two lemmas and Th. 9.1 b), the proof can be reduced to the σ-compact case (actually the compact case).

Lemma 10.1 Countable product of metric spaces is metrizable, namely its topology can be defined by a metric.

Proof. Let (X,τ) be the product topological space of metric spaces $\{(X_n,d_n)\}$. We can assume $\delta_n(X_n)\leqq 1$. For $x=(x_n)$ and $y=(y_n)$, we shall define a metric by:

$$(10.6) \qquad d(x,y) = \sum_{n=1}^{\infty} 2^{-n} d_n(x_n,y_n).$$

We shall show that the product topology τ is defined by the metric d.

First, $d(x,y)<2^{-n}\varepsilon$ implies $d_k(x_k,y_k)<\varepsilon$ for $k=1,2,\cdots,n$. On the other hand, for a given $\varepsilon>0$, choosing n such that $2^{-n}<\varepsilon/2$, we see that $d_k(x_k,y_k)<\varepsilon/2$ for $k=1,2,\cdots,n$ implies $d(x,y)<\varepsilon/2+\sum_{k=n+1}^{\infty} 2^{-k}=\varepsilon/2+2^{-n}<\varepsilon$. This completes the proof.

Lemma 10.2. Let (X,d) be a complete metric space, and φ be a homeomorphism which maps X into a topological space (Z,τ). Then, the image $\varphi(X)$ is a Borel set in Z. More strongly, $\varphi(X)$ is an intersection of a closed set and a G_δ-set (=countable intersection of open sets).

Proof. Put

$$(10.7) \qquad G_n = \{z\in Z \mid \exists O: \text{ open in } Z,\ z\in O,\ \varphi^{-1}(O)\neq\emptyset,\ \delta(\varphi^{-1}(O))<1/n\}.$$

Evidently G_n is open in Z, and contains $\varphi(X)$. We shall prove

$$(10.8) \qquad \varphi(X) = \overline{\varphi(X)} \cap \bigcap_{n=1}^{\infty} G_n .$$

Take an arbitrary element $z \in \overline{\varphi(X)} \cap \bigcap_{n=1}^{\infty} G_n$. For every n, there exists O_n such that $z \in O_n$, and $\delta(\varphi^{-1}(O_n))<1/n$. We can assume that $\{O_n\}$ is decreasing, hence $\{\varphi^{-1}(O_n)\}$ is decreasing. Take $x_n \in \varphi^{-1}(O_n)$, then $\{x_n\}$ becomes a Cauchy sequence in X, so that it converges to some $x \in X$. The limit point x does not depend on the choice of $\{x_n\}$. Since φ is continuous, $\varphi(x_n)$ tends to $\varphi(x)$.

We shall prove $z=\varphi(x)$. If $z \neq \varphi(x)$, there would exist disjoint neighbourhoods U of z and V of $\varphi(x)$. If we choose $x_n \in \varphi^{-1}(O_n \cap U)$, we would have $\varphi(x_n) \notin V$, so $\varphi(x_n)$ could not tend to $\varphi(x)$. Thus we have proved $z=\varphi(x) \in \varphi(X)$.

(q.e.d.)

§11. Baire field

<u>Definition 11.1.</u> Let (X,τ) be a topological space, and let $\mathbb{B}$ be the smallest σ-algebra in which all continuous functions are measurable. $\mathbb{B}$ is called the Baire field, and a set belonging to $\mathbb{B}$ is called a Baire set. $(X, \mathbb{B})$ is called the Baire measurable space associated with τ, and a measure on $\mathbb{B}$ is called a Baire measure.

If we denote the family of all continuous functions on X with $C(X)$, and the Borel field on $\mathbb{R}$ with $\mathcal{B}_0$, then we have

$$(11.1) \qquad \mathbb{B} = \mathcal{B}\Big(\bigcup_{f \in C(X)} f^{-1}(\mathcal{B}_0)\Big).$$

Since every continuous function is measurable with respect to the Borel field $\mathcal{B}$ of (X,τ), we have $\mathbb{B} \subset \mathcal{B}$.

If every open set is a Baire set, we have $\mathbb{B}=\mathcal{B}$. Especially for a metric space, we have $\mathbb{B}=\mathcal{B}$, because for every open set O, $d(x,O^c)$ is a continuous function of x and $O= \{x \in X \mid d(x,O^c)>0\}$.

If some point x of X can not be expressed as a countable intersection of its neighbourhoods, then $\{x\}$ does not belong to $\mathbb{B}$, so we have $\mathbb{B} \subsetneqq \mathcal{B}$. An example of such a space is the non-countable product space, assuming each $X^{(\lambda)}$ has more than two points. Another example is given as follows: Let Ω be the first non-countable ordinal number. The well-ordered set $[0,\Omega]$ becomes a topological space, if we define neighbourhoods of x by $(\alpha,x]$ $(\alpha<x\leqq\Omega)$. Then, every countable intersection of neighbourhoods of Ω is again a neighbourhood of Ω.

<u>Definition 11.2.</u> A topological space (X,τ) is said to be locally compact, if every point has a compact neighbourhood, namely if $\forall x \in X$, $\exists U$: open in X, $x \in U$ and $\overline{U}$ is compact.

<u>Theorem 11.1.</u> Let (X,τ) be a σ-compact and locally compact space. Then, the Baire measurable space $(X, \mathbb{B})$ is compact regular in τ.

<u>Remark.</u> A compact space is σ-compact and locally compact. Since there exists a non-metrizable compact space, this theorem is independent of Th. 10.1.

<u>Proof</u> (of Th.). Let $\mathfrak{A}$ be the family of all Baire sets B which satisfy:

$$\inf\{d^*_\mu(K,B) \mid K \text{ is compact and } K\subset B\} = 0. \tag{11.2}$$

Since $\mathfrak{A}$ is closed under countable intersection and countable union, in order to prove $\mathfrak{A} = \mathbb{B}$, it is sufficient to show that $f^{-1}(O)$ and $f^{-1}(F)$ belong to $\mathfrak{A}$, where $f \in C(X)$ and O

(resp. F) is an open (resp. a closed) set in $\mathbb{R}$.

Since $\mathbb{R}$ is a metric space, O can be expressed as a countable union of closed sets, and $O=\bigcup_{n=1}^{\infty} F_n$ implies $f^{-1}(O)=\bigcup_{n=1}^{\infty} f^{-1}(F_n)$. So, we need to show only $f^{-1}(F)\in \mathfrak{A}$. Since (X, τ) is σ-compact, we have $X=\bigcup_{n=1}^{\infty} K_n$. So we have

$$(11.3) \qquad f^{-1}(F) = \bigcup_{n=1}^{\infty} (f^{-1}(F)\cap K_n).$$

Evidently, every compact Baire set belongs to $\mathfrak{A}$. Therefore, if we prove that we can choose K_n from $\mathbb{B}$, then (11.3) assures $f^{-1}(F)\in \mathfrak{A}$.

<u>Lemma 11.1.</u> In a locally compact space X, a compact set is always contained in a compact Baire set.

<u>Proof.</u> Let K be a compact set, and U(x) be an open neighbourhood of x such that $\overline{U(x)}$ is compact. Since K is compact, K is covered by finite number of U(x). Thus, denoting this finite union with O, we have

$$(11.4) \qquad K\subset O, \quad O \text{ is open and } \overline{O} \text{ is compact.}$$

Here, we shall use the following lemma without proof.

<u>Lemma 11.2.</u> A compact space X is a normal space, namely for disjoint closed sets F_1 and F_2, there exists a continuous function f on X such that f=1 on F_1 and f=0 on F_2.

<u>Proof</u> can be found in any book on general topology, e.g. N. Bourbaki "Topologie générale" Chap. 9, §4.

Since $\overline{O}$ is compact, there exists a continuous function f on $\overline{O}$ such that f=1 on K and f=0 on $\overline{O}\cap O^c$ (=boundary of O). If we put f(x)=0 for $x\in \overline{O}^c$, we get a continuous function f on X. Evidently $f^{-1}(1)$ contains K and is contained in O, so that it is a compact Baire set containing K. (q.e.d.)

Remark. If we apply Lemma 11.2 to $F_1=\{x\}$ and $F_2=U(x)^c$, we see that in a locally compact space, every point has a fundamental system of neighbourhoods consisting of Baire sets.

Here we shall mention some facts concerning the relation between the product of topologies and that of σ-algebras.

Theorem 11.2. Let $\{(X^{(\lambda)},\tau^{(\lambda)})\}_{\lambda\in\Lambda}$ be a family of topological spaces, and (X,τ) be its product topological space. Let $\mathcal{B}^{(\lambda)}$ and $\mathbb{B}^{(\lambda)}$ be the Borel and Baire field of $(X^{(\lambda)},\tau^{(\lambda)})$, and $\mathcal{B}$ and $\mathbb{B}$ be those of (X,τ). Finally, let $(X,\mathcal{B}')$ and $(X,\mathbb{B}')$ be the product measurable space of $\{(X^{(\lambda)},\mathcal{B}^{(\lambda)})\}$ and $\{(X^{(\lambda)},\mathbb{B}^{(\lambda)})\}$. Then we have:

a) $\mathbb{B}'\subset\mathcal{B}'$, $\mathbb{B}'\subset\mathbb{B}$, $\mathcal{B}'\subset\mathcal{B}$.

b) If each $(X^{(\lambda)},\tau^{(\lambda)})$ is compact, then we have $\mathbb{B}'=\mathbb{B}$. (The result is valid even if finite number of $(X^{(\lambda)},\tau^{(\lambda)})$ are σ-compact and locally compact).

c) If each $(X^{(\lambda)},\tau^{(\lambda)})$ is a separable metric space, and if Λ is countable, then we have $\mathcal{B}'=\mathcal{B}$.

d) If Λ is non-countable, we have $\mathcal{B}'\subsetneqq\mathcal{B}$. (We assume that each $X^{(\lambda)}$ has more than two points).

Proof. a) Since $\mathbb{B}'=\mathcal{B}(\bigcup_{\lambda\in\Lambda}p^{(\lambda)-1}(\mathbb{B}^{(\lambda)}))$ and $\mathcal{B}'=\mathcal{B}(\bigcup_{\lambda\in\Lambda}p^{(\lambda)-1}(\mathcal{B}^{(\lambda)}))$, $\mathbb{B}^{(\lambda)}\subset\mathcal{B}^{(\lambda)}$ implies $\mathbb{B}'\subset\mathcal{B}'$. Since every projection is continuous in the product topology, it is measurable with respect to the Borel fields, so that $\mathcal{B}'\subset\mathcal{B}$. For every continuous function f on $X^{(\lambda)}$, $f\circ p^{(\lambda)}$ is a continuous function on X, therefore $\mathbb{B}'\subset\mathbb{B}$.

b) We shall prove that for $f \in C(X)$ and for an open set O in $\mathbb{R}$, $f^{-1}(O)$ belongs to $\mathbb{B}'$. Expressing $O = \bigcup_{n=1}^{\infty} F_n$, we have $f^{-1}(O) = \bigcup_{n=1}^{\infty} f^{-1}(F_n)$. So, it is sufficient to prove that:

$$(11.5) \qquad \exists B_n \in \mathbb{B}', \quad f^{-1}(F_n) \subset B_n \subset f^{-1}(O).$$

From the assumption, (X,τ) is compact (resp. σ-compact and locally compact), so $f^{-1}(F_n)$ is a compact set (resp. a countable union of compact sets). From the definition of the product topology and from the remark before this theorem, every point of (X,τ) has a fundamental system of neighbourhoods belonging to $\mathbb{B}'$. So, covering a compact set by finite number of such $\mathbb{B}'$-measurable neighbourhoods, we have (11.5).

c) Since (X,τ) is a separable metric space, every open set O in (X,τ) can be expressed as a countable union of open balls. Take the metric d defined by (10.6), then $d(x,x_0)$ is evidently a $\mathcal{B}'$-measurable function of x, so that such open balls belong to $\mathcal{B}'$.

d) Since we are assuming that the topology is Hausdorff, every one-point set is a Borel set. However as seen in §8, every $\mathcal{B}'$-measurable set belongs to some $p_M^{-1}(\mathcal{B}_M)$, M being a countable subset of Λ, so it is not a one-point set. (q.e.d.)

§12. Product measure

In this section, we shall study an important extension theorem which is independent of compact regularity. It claims the existence of the product measure, without any condition on measurable spaces.

First, as preliminary discussions, we shall review on the

finite product. Let $(X_1, \mathcal{B}_1)$ and $(X_2, \mathcal{B}_2)$ be measurable spaces, and $(X_1 \times X_2, \mathcal{B}_1 \times \mathcal{B}_2)$ be their product measurable space, namely:

$$(12.1) \qquad \mathcal{B}_1 \times \mathcal{B}_2 = \mathcal{B}(p_1^{-1}(\mathcal{B}_1) \cup p_2^{-1}(\mathcal{B}_2)).$$

For every $E \in \mathcal{B}_1 \times \mathcal{B}_2$, its section $E(x_1)$ belongs to $\mathcal{B}_2$, where

$$(12.2) \qquad E(x_1) = \{x_2 \in X_2 \mid (x_1, x_2) \in E\}.$$

Let μ_1 (resp. μ_2) be a measure on $\mathcal{B}_1$ (resp. $\mathcal{B}_2$), then for every $E \in \mathcal{B}_1 \times \mathcal{B}_2$, $\mu_2(E(x_1))$ is a $\mathcal{B}_1$-measurable function of x_1, and we can define the product measure $\mu_1 \times \mu_2$ by:

$$(12.3) \qquad (\mu_1 \times \mu_2)(E) = \int_{X_1} \mu_2(E(x_1))\, d\mu_1(x_1).$$

If $E_1 \in \mathcal{B}_1$ and $E_2 \in \mathcal{B}_2$, then we have

$$(12.4) \qquad (\mu_1 \times \mu_2)(E_1 \times E_2) = \mu_1(E_1)\mu_2(E_2).$$

The product is commutative in the following sense: For the map $\varphi: (x_1, x_2) \to (x_2, x_1)$, we have $(\mu_2 \times \mu_1)(\varphi(E)) = (\mu_1 \times \mu_2)(E)$.

The product is also associative, namely we have $(\mu_1 \times \mu_2) \times \mu_3 = \mu_1 \times (\mu_2 \times \mu_3)$ on $(X_1 \times X_2 \times X_3, \mathcal{B}_1 \times \mathcal{B}_2 \times \mathcal{B}_3)$. This common measure is denoted with $\mu_1 \times \mu_2 \times \mu_3$. By mathematical induction, we can define any finite product: $(\mu_1 \times \mu_2 \times \cdots \times \mu_p) \times (\mu_{p+1} \times \mu_{p+2} \times \cdots \times \mu_n)$ is independent of p, and is denoted with $\mu_1 \times \mu_2 \times \cdots \times \mu_n$ (or $\prod_{k=1}^{n} \mu_k$).

Now, suppose that $\{(X^{(\lambda)}, \mathcal{B}^{(\lambda)})\}_{\lambda \in \Lambda}$ is a family of measurable spaces and that $\mu^{(\lambda)}$ is a probability measure on $(X^{(\lambda)}, \mathcal{B}^{(\lambda)})$. On every finite product measurable space $(X_L, \mathcal{B}_L)$, we can consider the finite product measure $\mu_L = \prod_{\lambda \in L} \mu^{(\lambda)}$. (Since the product is commutative, μ_L depends only on the set L, not depending on the choice of the order of $\{\lambda_k\}=L$). $\{\mu_L\}$ becomes a self-consistent family, because for $L \subset L'$ we have

$$p_{LL'} \circ \mu_{L'}(B_L) = \mu_{L'}(B_L \times X_{L'-L}) = (\mu_L \times \mu_{L'-L})(B_L \times X_{L'-L})$$
$$= \mu_L(B_L)\mu_{L'-L}(X_{L'-L}) = \mu_L(B_L).$$

Theorem 12.1. Let $\{(X^{(\lambda)}, \mathcal{B}^{(\lambda)}, \mu^{(\lambda)})\}_{\lambda \in \Lambda}$ be a family of probability measure spaces. The family $\{\mu_L\}$ of finite product measures is extendable to a σ-additive measure μ on the product measurable space $(X, \mathcal{B})$, without any condition on $(X^{(\lambda)}, \mathcal{B}^{(\lambda)})$.

Definition 12.1. The measure μ given in Th. 12.1 is called the product measure of $\mu^{(\lambda)}$, and denoted with $\prod_{\lambda \in \Lambda} \mu^{(\lambda)}$.

Proof. By Th. 8.1, it is sufficient to prove the theorem only for $\Lambda=\{1,2,\cdots\}$. Denote the product of $\{(X^{(k)}, \mathcal{B}^{(k)}, \mu^{(k)})\}_{k=1,2,\cdots,n}$ with $(X_n, \mathcal{B}_n, \mu_n)$, and the product of $\{(X^{(k)}, \mathcal{B}^{(k)}, \mu^{(k)})\}_{k=m+1,m+2,\cdots,n}$ with $(X_{mn}, \mathcal{B}_{mn}, \mu_{mn})$. Then, $\{\mu_n\}$ is a self-consistent sequence and induces a finitely additive measure μ on $\mathcal{F} = \bigcup_{n=1}^{\infty} p_n^{-1}(\mathcal{B}_n)$. In order to show that μ can be extended to a σ-additive measure on $\mathcal{B}$, we shall check the Hopf's condition: For every decreasing sequence $\{F_n\}$ in $\mathcal{F}$,

$$\text{(12.5)} \qquad \lim_{n\to\infty} \mu(F_n) > 0 \quad \text{implies} \quad \bigcap_{n=1}^{\infty} F_n \neq \emptyset.$$

As explained in the proof of Th. 6.1, it is sufficient to prove (12.5) under the assumption: $\forall n,\ F_n \in p_n^{-1}(\mathcal{B}_n)$. Then we have $\exists B_n$, $F_n = p_n^{-1}(B_n)$. Since $\{F_n\}$ is decreasing, we have

(12.6) $\quad m<n \Longrightarrow B_n \subset p_{mn}^{-1}(B_m) = B_m \times X_{mn}.$

The assumption in (12.5) is equivalent with

(12.7) $\quad \lim_{n\to\infty} \mu_n(B_n) > 0.$

From (12.6), we see that for every $x^{(1)} \in X^{(1)}$,

(12.6)' $\quad 2 \leqq m<n \Longrightarrow B_n(x^{(1)}) \subset B_m(X^{(1)}) \times X_{mn}.$

On the other hand, since $\mu_n = \mu_1 \times \mu_{1n}$, we have

$$\mu_n(B_n) = \int_{X^{(1)}} \mu_{1n}(B_n(x^{(1)}))\, d\mu_1(x^{(1)}).$$

Since the integrand is decreasing in n, Lebesgue's convergence theorem assures:

$$\lim_{n\to\infty} \mu_n(B_n) = \int_{X^{(1)}} \lim_{n\to\infty} \mu_{1n}(B_n(x^{(1)}))\, d\mu_1(x^{(1)}).$$

Combining with (12.7), we see that

(12.7)' $\quad \exists x^{(1)} \in X^{(1)}, \quad \lim_{n\to\infty} \mu_{1n}(B_n(x^{(1)})) > 0.$

(12.6)' and (12.7)' are the same conditions with (12.6) and (12.7), if we replace B_n by $B_n(x^{(1)})$ and μ_n by μ_{1n}. Therefore applying the same discussions, we can find some $x^{(2)} \in X^{(2)}$ such that $\lim_{n\to\infty} \mu_{2n}(B_n(x^{(1)}, x^{(2)}))>0$. By mathematical induction, we shall define $x^{(m)}$ step by step, then we have a sequence $x=(x^{(m)})$ such that

(12.8) $\quad \forall m, \quad \lim_{n\to\infty} \mu_{mn}(B_n(x^{(1)}, x^{(2)}, \cdots, x^{(m)})) > 0.$

Especially we have $B_n(x^{(1)},x^{(2)},\cdots,x^{(m)})\neq\emptyset$, so that $(x^{(1)}, x^{(2)},\cdots,x^{(m)})$ must belong to B_m. Thus we have $p_m(x)\in B_m$, hence $x\in p_m^{-1}(B_m)$. Therefore $x\in \bigcap_{m=1}^{\infty} p_m^{-1}(B_m)=\bigcap_{m=1}^{\infty} F_m$.

(q.e.d.)

§13. Suslin set, Luzin set

This section is devoted to preliminary discussions for the next section.

Theorem 13.1. Let (X,d) be a complete separable metric space. Whenever X is non-countable, (X,d) contains a subspace which is homeomorphic to $\{0,1\}^\infty$, hence X has the power of continuum. Here, $\{0,1\}^\infty$ is the countable product of two-point spaces.

Proof. A point $x\in X$ is called a condensation point, if every neighbourhood of x is non-countable. Let A be the set of all condensation points, then the subspace (A,d) is complete and has no isolated point as shown below: First, from the definition of a condensation point and the separability of (X,d), A^c becomes a countable union of open countable sets, thus A^c is an open countable set. So that A is closed, hence (A,d) is complete. From the definition of a condensation point, every neighbourhood $U(x)$ of $x\in A$ is non-countable, so that $U(x)\cap A$ is non-countable because A^c is countable. Thus any $x\in A$ is not an isolated point of A.

Replacing X by A, if necessary, we can assume that X has no isolated point. Take two open balls U_0 and U_1 such that $\bar{U}_0\cap\bar{U}_1=\emptyset$. Since U_0(resp. U_1) has more than two points, we can take two open balls U_{00} and U_{01} in U_0(resp. U_{10} and U_{11} in U_1) such that $\bar{U}_{00}\cap\bar{U}_{01}=\emptyset$(resp. $\bar{U}_{10}\cap\bar{U}_{11}=\emptyset$).

Repeating this procedure, we shall get open balls $\{U_{\alpha_1\alpha_2\cdots\alpha_n}\}$, $\alpha_1,\alpha_2,\cdots,\alpha_n=0$ or 1 such that:

1) For fixed n, $\bar{U}_{\alpha_1\alpha_2\cdots\alpha_n}$ are mutually disjoint,

2) $U_{\alpha_1\alpha_2\cdots\alpha_n\alpha_{n+1}} \subset U_{\alpha_1\alpha_2\cdots\alpha_n}$.

Here, we can assume that the radius of $U_{\alpha_1\alpha_2\cdots\alpha_n}$ is less than 2^{-n}.

For every $\alpha=(\alpha_n)\in\{0,1\}^\infty$, the sequence $\{\bar{U}_{\alpha_1\alpha_2\cdots\alpha_n}\}$ is a decreasing sequence of non-empty closed sets whose diameters tend to zero. Since (X,d) is complete, $\bigcap_{n=1}^{\infty}\bar{U}_{\alpha_1\alpha_2\cdots\alpha_n}$ is just a one-point set $\{x_\alpha\}$. We shall denote the map: $\{0,1\}^\infty \ni \alpha \to x_\alpha \in X$ with Φ.

Φ is injective because of 1). Φ is continuous because $\alpha_k'=\alpha_k$ for $1\leq k\leq n$ implies $x_{\alpha'}\in\bar{U}_{\alpha_1\alpha_2\cdots\alpha_n}$. So, Φ must be a homeomorphism because $\{0,1\}^\infty$ is compact. Therefore, denoting the image $\Phi(\{0,1\}^\infty)$ with B, we get $B\cong\{0,1\}^\infty$.

Definition 13.1. Let (X,τ) be a topological space. A subset Y of X is called a Luzin set (resp. Suslin set), if Y is a continuous injective (resp. a continuous) image of a complete separable metric space, namely if $Y=f(Z)$ for some complete separable metric space (Z,d) and some continuous injective (resp. continuous) map f from Z into X. (X,τ) is called a Luzin (resp. Suslin) space, if X itself is a Luzin (resp. Suslin) set.

A Luzin space is nothing else but a topological space whose topology is weaker than some complete separable metric topology.

If (X_k,τ_k) are Luzin (resp. Suslin) spaces, then their countable product (X,τ) is a Luzin (resp. Suslin) space, because: $X_k=f_k(Z_k)$ imply $X=f(\prod_{k=1}^{\infty} Z_k)$ where $f: (z_k)\to(f_k(z_k))$: If each Z_k is a complete separable metric space, then $\prod_{k=1}^{\infty} Z_k$ is so in the product topology, and if each f_k is continuous, then f is so in the product topologies.

If (X_k,τ_k) are Luzin (resp. Suslin) spaces, then their disjoint countable union is a Luzin (resp. Suslin) space in the sum topology, because: $X_k=f_k(Z_k)$ imply $X=f(\dot{\bigcup}_{k=1}^{\infty} Z_k)$ where $f=f_k$ on Z_k. If each Z_k is a complete separable metric space, then their disjoint union is so in the sum topology, and if each f_k is continuous, then f is so in the sum topologies.

<u>Theorem 13.2.</u> In a Luzin (resp. Suslin) space (X,τ), every Borel set is a Luzin (resp. Suslin) set.

<u>Proof.</u> Let $\mathfrak{A}$ be the family of all Luzin (resp. Suslin) sets in X, and we shall prove that $\mathfrak{A}$ contains the Borel field. Suppose that $X=f(Z)$.

First, note that in a complete separable metric space (Z, d), a closed set is evidently complete. An open set O may not be complete in the metric d, but it is complete in an equivalent metric $\tilde{d}$:

$$\tilde{d}(x,y) = d(x,y) + \left|\frac{1}{d(x,O^c)} - \frac{1}{d(y,O^c)}\right|. \tag{13.1}$$

In the space (X,τ), a closed set F belongs to $\mathfrak{A}$, because it is the image of $f^{-1}(F)$, which is closed in Z. An open set O in X belongs to $\mathfrak{A}$, because it is the image of $f^{-1}(O)$, which is open in Z.

$\mathfrak{A}$ is closed under countable intersection as shown below: Suppose $E_k \in \mathfrak{A}$, then $\prod_{k=1}^{\infty} E_k$ is a Luzin (resp. Suslin) set in

X^∞. Since the diagonal Δ is closed in X^∞, $\prod_{k=1}^{\infty} E_k \cap \Delta$ is a Luzin (resp. Suslin) set, so that its continuous image by p_1, namely $\bigcap_{k=1}^{\infty} E_k$ is a Luzin (resp. Suslin) set.

$\mathcal{A}$ is closed under disjoint countable union as shown below: Suppose $E_k \in \mathcal{A}$ and that $\{E_k\}$ is mutually disjoint, then $\bigcup_{k=1}^{\infty} E_k$ is a Luzin (resp. Suslin) set in the sum topology. Since the sum topology is stronger than τ, $\bigcup_{k=1}^{\infty} E_k$ is a Luzin (resp. Suslin) set also in τ.

Put $\mathcal{B}' = \mathcal{A} \cap \mathcal{A}^c$, then $\mathcal{B}'$ is a σ-algebra and contains all open sets, therefore $\mathcal{B}'$ contains all Borel sets.

(q.e.d.)

<u>Lemma 13.1.</u> In a topological space (X,τ), mutually disjoint two Suslin sets S and S' are separated by Borel sets, namely $^\exists B \supset S$, $^\exists B' \supset S'$, B and B' are Borel sets and $B \cap B' = \emptyset$.

<u>Corollary</u>. If both S and S^c are Suslin sets, then S is a Borel set.

<u>Proof</u> (of Lemma). First, note that if S_n and S'_m are separated by B_{nm} and B'_{mn} for each pair (n,m), then $S = \bigcup_{n=1}^{\infty} S_n$ and $S' = \bigcup_{m=1}^{\infty} S'_m$ are separated by B and B', where $B = \bigcup_{n=1}^{\infty} (\bigcap_{m=1}^{\infty} B_{nm})$ and $B' = \bigcup_{m=1}^{\infty} (\bigcap_{n=1}^{\infty} B'_{mn})$. In other words, if S and S' are not separated by Borel sets, then some S_n and S'_m are not separated by Borel sets.

Suppose that $S=f(Z)$ and $S'=f'(Z')$. Take a countable open covering of Z (resp. Z') by sets of diameters $\leqq 1/2$. If S and S' can not be separated by Borel sets, then there exists an open set U_1 in Z (resp. U'_1 in Z') of the diameter $\leqq 1/2$ such that $f(U_1)$ and $f'(U'_1)$ can not be separated by Borel sets. Again some $f(U_2)$ and $f'(U'_2)$ can not be separated by Borel sets, U_2 (resp. U_2') being an open set which is contained in U_1 (resp. U_1') and of the diameter$\leqq 1/4$. Repeating this procedure, we can find a decreasing sequence $\{U_n\}$ (resp. $\{U_n'\}$) such that $f(U_n)$ and $f'(U_n')$ can not be separated by Borel sets for any n. Since $\{U_n\}$ is a decreasing sequence of non-empty sets whose diameters tend to zero, $\bigcap_{n=1}^{\infty} \overline{U}_n$ becomes a one-point set $\{z\}$. Similarly we have $\bigcap_{n=1}^{\infty} \overline{U}_n'=\{z'\}$.

Since f and f' are continuous, $f(U_n)$ tends to $f(z)$ while $f'(U_n')$ tends to $f'(z')$. Therefore, for open neighbourhoods V of $f(z)$ and V' of $f'(z')$ satisfying $V \cap V'=\emptyset$, if n is large enough, we have $f(U_n) \subset V$ and $f'(U_n') \subset V'$. This is a contradiction to the construction of $\{U_n\}$ and $\{U_n'\}$.

(q.e.d.)

Remark. This lemma can be applied to countable number of disjoint Suslin sets. The proof is easily reduced to the case of two Suslin sets.

Theorem 13.3. In a topological space (X,τ), every Luzin set is a Borel set.

Corollary. In a Luzin space, a Luzin set is nothing else but a Borel set.

Proof (of Th.). Let L be a Luzin set and suppose that $L=f(Z)$. Let $\{Z_{n_1}\}_{n_1=1,2,\cdots}$ be a countable partition of Z by Borel sets of the diameters$\leqq 1/2$, $\{Z_{n_1 n_2}\}_{n_2=1,2,\cdots}$ be a countable partition of Z_{n_1} by Borel sets of the diameters$\leqq 1/4$, and so on. Then, $\{f(Z_{n_1})\}_{n_1=1,2,\cdots}$ is a partition of L, $\{f(Z_{n_1 n_2})\}_{n_2=1,2,\cdots}$ is a partition of $f(Z_{n_1})$, and so on.

Since Z_{n_1} etc. is a Borel set in Z, it is a Luzin set by Th. 13.2, so that $f(Z_{n_1})$ etc. is a Luzin set in X. Since $\{f(Z_{n_1})\}$ is mutually disjoint, by Lemma 13.1 and its remark, there exists a family $\{B_{n_1}\}$ of mutually disjoint Borel sets such that $B_{n_1} \supset f(Z_{n_1})$. Here, we can assume that $B_{n_1} \subset \overline{f(Z_{n_1})}$.

Repeating this procedure, we can find Borel sets $B_{n_1 n_2 \cdots n_r}$ such that:

1) $f(Z_{n_1 n_2 \cdots n_r}) \subset B_{n_1 n_2 \cdots n_r} \subset \overline{f(Z_{n_1 n_2 \cdots n_r})}$,

2) for fixed r, $B_{n_1 n_2 \cdots n_r}$ are mutually disjoint,

3) $B_{n_1 n_2 \cdots n_r n_{r+1}} \subset B_{n_1 n_2 \cdots n_r}$.

Put $B = \bigcap_{r=1}^{\infty} \left(\bigcup_{n_1, n_2, \cdots, n_r} B_{n_1 n_2 \cdots n_r} \right)$, then B is a Borel set in X. We shall prove $L=B$. Evidently we have $L \subset B$. Conversely, take an arbitrary element $x \in B$, then we have $\forall r, \exists n_1, n_2, \cdots, n_r$, $x \in B_{n_1 n_2 \cdots n_r}$. However from 2) and 3), such an $(n_1, n_2, \cdots, n_r)$ is unique and consistent. The corresponding $\{Z_{n_1 n_2 \cdots n_r}\}_{r=1,2,\cdots}$ becomes a decreasing sequence of non-empty sets whose diameters tend to zero, so $\bigcap_{r=1}^{\infty} \overline{Z}_{n_1 n_2 \cdots n_r}$ is a one-point set $\{z\}$. We shall prove $x=f(z)$. If $x \neq f(z)$, there would exist disjoint neighbourhoods U of x and V of $f(z)$. For sufficiently large r, we would have $f(Z_{n_1 n_2 \cdots n_r}) \subset V$, hence $B_{n_1 n_2 \cdots n_r} \subset \overline{f(Z_{n_1 n_2 \cdots n_r})} \subset U^c$. This contradicts to $x \in B_{n_1 n_2 \cdots n_r}$. Thus we get $x = f(z) \in L$. (q.e.d.)

Remark. A Suslin set is not always a Borel set, even in a complete separable metric space. We can find a counter example for the space $\mathbf{N}^{\infty}$ in N. Bourbaki "Topologie générale" Chap. 9,

§6, Exercise 6. Another example is known for the space C(I) of all continuous functions on a compact interval I with the uniform norm. The set of all differentiable functions on I is the complement of a Suslin set, but not a Borel set in C(I).

§14. Standard measurable space

Definition 14.1. A measurable space $(X, \mathcal{B})$ is said to be standard, if it is isomorphic to a Borel measurable space of a complete separable metric space.

A standard measurable space is worth studying because of Corollary of Th. 10.1. (The Borel measurable space of a σ-compact metric space is standard, as seen later in Th. 14.3). First we shall study the isomorphism theorem, and next some criterions for standardness.

We shall start with general discussions on measurable isomorphisms. When $(X, \mathcal{B})$ and $(X', \mathcal{B}')$ are isomorphic by a measurable isomorphism φ, we shall denote it with $(X, \mathcal{B}) \underset{\varphi}{\sim} (X', \mathcal{B}')$.

Theorem 14.1. a) If $(X, \mathcal{B}) \underset{\varphi}{\sim} (X', \mathcal{B}')$, then for any subset Y of X, we have $(Y, \mathcal{B} \cap Y) \underset{\varphi}{\sim} (\varphi(Y), \mathcal{B}' \cap \varphi(Y))$.

b) Suppose that $\{Y_n\}$ (resp. $\{Y'_n\}$) is a countable partition of X (resp. X') and that each Y_n (resp. Y'_n) belongs to $\mathcal{B}$ (resp. $\mathcal{B}'$). Then, $(Y_n, \mathcal{B} \cap Y_n) \underset{\varphi_n}{\sim} (Y'_n, \mathcal{B}' \cap Y'_n)$ for every n implies $(X, \mathcal{B}) \sim (X', \mathcal{B}')$.

c) $(X^{(\lambda)}, \mathcal{B}^{(\lambda)}) \underset{\varphi(\lambda)}{\sim} (X'^{(\lambda)}, \mathcal{B}'^{(\lambda)})$ for every λ implies that their product measurable spaces are isomorphic.

d) Suppose that $(X, \mathcal{B})$ is the Borel measurable space of a topology τ. $(X', \mathcal{B}')$ is isomorphic to $(X, \mathcal{B})$, if and only if there exists a topology τ' on X' such that (X,τ) and (X',τ') are homeomorphic and $\mathcal{B}'$ is the Borel field of τ'.

e) Let $(X,\mathcal{B})$ and $(X',\mathcal{B}')$ be measurable spaces. If $(X,\mathcal{B})\sim(Y',\mathcal{B}'\cap Y')$ and $(X',\mathcal{B}')\sim(Y,\mathcal{B}\cap Y)$ for some $Y\in\mathcal{B}$ and $Y'\in\mathcal{B}'$, then we have $(X,\mathcal{B})\sim(X',\mathcal{B}')$.

<u>Proof.</u> a) $\varphi^{-1}(\mathcal{B}'\cap\varphi(Y))=\varphi^{-1}(\mathcal{B}')\cap Y=\mathcal{B}\cap Y$ assures our theorem.

b) Consider the map $\varphi: X\to X'$ such that $\varphi=\varphi_n$ on Y_n. For every $B\in\mathcal{B}$, we have $\varphi(B)=\bigcup_{n=1}^{\infty}\varphi_n(B\cap Y_n)\in\mathcal{B}'$. Combining it with a similar result for φ^{-1}, we see that φ is a measurable isomorphism.

c) Denote the product measurable spaces with $(X,\mathcal{B})$ and $(X',\mathcal{B}')$. Consider the map $\varphi: X\to X'$ such that $p^{(\lambda)}\circ\varphi=\varphi^{(\lambda)}\circ p^{(\lambda)}$, then for every $B'^{(\lambda)}\in\mathcal{B}'^{(\lambda)}$ we have $\varphi^{-1}(p^{(\lambda)-1}(B'^{(\lambda)}))=p^{(\lambda)-1}(\varphi^{(\lambda)-1}(B'^{(\lambda)}))\in\mathcal{B}$. Thus we get $\varphi^{-1}(\mathcal{B}')\subset\mathcal{B}$. Combining it with a similar result for φ, we see that φ is a measurable isomorphism.

d) Since every homeomorphism is a measurable isomorphism with respect to Borel fields, the condition is sufficient. Suppose $(X,\mathcal{B})\underset{\varphi}{\sim}(X',\mathcal{B}')$, and consider the image topology $\varphi(\tau)$. Evidently (X,τ) and $(X',\varphi(\tau))$ are homeomorphic. If we denote the Borel field of $\varphi(\tau)$ with $\mathcal{B}''$, then we have $\varphi(\mathcal{B})=\mathcal{B}''$. Combining it with $\varphi(\mathcal{B})=\mathcal{B}'$, we get $\mathcal{B}'=\mathcal{B}''$.

e) Suppose $(X,\mathcal{B})\underset{\varphi}{\sim}(Y',\mathcal{B}'\cap Y')$ and $(X',\mathcal{B}')\underset{\psi}{\sim}(Y,\mathcal{B}\cap Y)$. Put $Y_0=X$ and $Y_0'=X'$, and by mathematical induction put $Y_{n+1}=\psi(Y_n')$ and $Y_{n+1}'=\varphi(Y_n)$. Since $\{Y_n\}$ is a decreasing sequence of $\mathcal{B}$-measurable sets, putting $Z_n=Y_n-Y_{n+1}$, we have a countable partition $\{Z_n\}$ of X together with $Z_\infty=\bigcap_{n=1}^{\infty}Y_n$. Similar discussions can be applied also to X'. By a), we get $(Z_{2n}, \mathcal{B}\cap Z_{2n})\underset{\varphi}{\sim}(Z'_{2n+1}, \mathcal{B}'\cap Z'_{2n+1})$, $(Z_{2n+1}, \mathcal{B}\cap Z_{2n+1})\underset{\psi^{-1}}{\sim}(Z'_{2n}, \mathcal{B}'\cap Z'_{2n})$ and $(Z_\infty, \mathcal{B}\cap Z_\infty)\underset{\varphi}{\sim}(Z'_\infty, \mathcal{B}'\cap Z'_\infty)$. Thus, by b) we get $(X,\mathcal{B})\sim(X',\mathcal{B}')$. (q.e.d.)

Theorem 14.2. Let $(X,\mathcal{B})$ be a standard measurable space. The cardinal number of X is finite, countable or the power of continuum. If X is finite or countable, then $\mathcal{B}$ is the Borel field of the discrete topology. If X has the power of continuum, $(X,\mathcal{B})$ is isomorphic to $(M, \mathcal{B}_M)$ where $M=\{0,1\}^\infty$ and $\mathcal{B}_M$ is the Borel field of the product topology. Namely, two standard measurable spaces are isomorphic, whenever they have the same cardinal number.

Proof. The statement on the cardinal number of X is true because of Th. 13.1. Since every one-point set belongs to $\mathcal{B}$, if X is finite or countable, every subset belongs to $\mathcal{B}$, thus $\mathcal{B}$ is the Borel field of the discrete topology.

Suppose that X has the power of continuum. Without loss of generality, we can assume that $(X,\mathcal{B})$ is the Borel measurable space of a complete separable metric space. By Th. 13.1, X has a subspace B which is homeomorphic to M, therefore we have $(M, \mathcal{B}_M)\sim(B, \mathcal{B}\cap B)$. So, by Th. 14.1 e), it is sufficient to prove that $(X,\mathcal{B})$ is isomorphic to some sub-measurable space of $(M, \mathcal{B}_M)$.

As seen in the proof of Th. 10.1, X is homeomorphic to some Borel set Y of $[0,1]^\infty$, therefore $(X,\mathcal{B})$ is isomorphic to $(Y, \mathcal{B}'\cap Y)$, where $([0,1]^\infty, \mathcal{B}')$ is the Borel measurable space of $[0,1]^\infty$. So, by Th. 14.1 a), it is sufficient to prove $([0,1]^\infty, \mathcal{B}')\sim(M, \mathcal{B}_M)$. First, the Borel measurable space of

[0,1] is isomorphic to $(M, \mathcal{B}_M)$ as shown below. Therefore by Th. 14.1 c), we have $([0,1]^\infty, \mathcal{B}') \sim (M, \mathcal{B}_M)^\infty$. However, since $M=\{0,1\}^\infty$, we get easily $(M, \mathcal{B}_M)^\infty \sim (M, \mathcal{B}_M)$ by changing the parameter of the product of $\{0,1\}$.

Now, we shall prove the suspended part: $([0,1], \mathcal{B}_1) \sim (M, \mathcal{B}_M)$, $\mathcal{B}_1$ being the Borel field of [0,1]. Let A be the set of all binary rational numbers in [0,1], and B be its complement. Every $x \in B$ has a unique binary expansion:

$$(14.1) \qquad x = \sum_{n=1}^{\infty} m_n(x) 2^{-n}, \qquad m_n(x)=0 \quad \text{or} \quad 1.$$

The map Φ: $x \to (m_n(x)) \in \{0,1\}^\infty$ is homeomorphic on B, because $m_n(x)=m_n(x')$ for $1 \leqq n \leqq N$ is equivalent to $\frac{\ell}{2^N} < x, x' < \frac{\ell+1}{2^N}$ for some ℓ. Since both A and $\Phi(B)^c$ are countable, we can consider a one-to-one correspondence between them, and regard Φ as a bijection from [0,1] to $\{0,1\}^\infty$. By Th. 14.1 b), Φ is a measurable isomorphism from $([0,1], \mathcal{B}_1)$ to $(M, \mathcal{B}_M)$.

(q.e.d.)

Next, we shall study some criterions for standardness.

<u>Theorem 14.3.</u> a) Let (X,τ) be a Luzin space, then the Borel measurable space $(X, \mathcal{B})$ is standard.

b) Let (X',τ') be another Luzin space, $(X', \mathcal{B}')$ be its Borel measurable space, and f be a measurable injection from $(X, \mathcal{B})$ into $(X', \mathcal{B}')$. Then, f(X) is a Luzin set and f^{-1} is measurable.

<u>Corollary.</u> Let (X,τ) be a Luzin space, and τ' be any topology which is weaker than τ. Then, the Borel fields of τ and τ' are identical.

<u>Proof</u> (of Th.). a) Since (X,τ) is a Luzin space, it is the

image of a complete separable metric space (Z,d) by a continuous bijection f. Then, f is a measurable isomorphism with respect to Borel fields as proved below: Every Borel set B in Z is a Luzin set by Th. 13.2, so that its continuous injective image $f(B)$ is a Luzin set, therefore $f(B)$ is a Borel set in X by Th. 13.3. This means that f^{-1} is Borel measurable.

b) is proved in a similar way. Since $(X', \mathcal{B}')$ is σ-separated, the graph $G=\{(x,f(x))|x \in X\}$ is a Borel set in $X\times X'$ in virtue of Th. 7.1. (The sufficiency part of Th. 7.1 is valid even if φ is not surjective). Therefore, for every $B\in\mathcal{B}$, $(B\times X')\cap G$ is a Borel set in $X\times X'$. Since $X\times X'$ is a Luzin space, $(B\times X')\cap G$ is a Luzin set, therefore its continuous injective image $p_2((B\times X')\cap G)=f(B)$ is a Luzin set, hence $f(B)$ is a Borel set in X'. This means that f^{-1} is Borel measurable. (q.e.d.)

Theorem 14.4. Assume that the Borel measurable space $(X,\mathcal{B})$ is standard, then (X,τ) must be a Luzin space if a) (X,τ) is a subspace of a Luzin space (X',τ'), or b) if (X,τ) satisfies the second axiom of countability, namely if X has a countalbe base $\{O_n\}$ of open sets.

Proof. a) is a result of Th. 14.3 b). If $(X,\mathcal{B})$ is standard, then there exists a Borel measurable bijection f from a complete separable metric space Z onto (X,τ), so that $X=f(Z)$ must be a Luzin set in X'.

We shall prove b). Denoting the indicator function of O_n with $c_n(x)$, the map Φ: $x\to(c_n(x))\in\{0,1\}^\infty=M$ is measurable and Φ^{-1} is continuous. Since $(X,\mathcal{B})$ is isomorphic to $(\Phi(X),$

$\mathcal{B}_M \cap \Phi(X))$, if $(X,\mathcal{B})$ is standard, then the latter is standard, so that $\Phi(X)$ is a Luzin set by a). Since $X=\Phi^{-1}(\Phi(X))$ is a continuous injective image of $\Phi(X)$, X is a Luzin space.

(q.e.d.)

But in general, even if (X,τ) is not a Luzin space, the Borel measurable space $(X,\mathcal{B})$ may be standard.

Example 14.1. Let $\mathbb{R}^n$ be the n-dimensional Euclidean space and $\mathcal{O}$ be the family of all open sets. Put

(14.2) $$\mathcal{O}_\tau = \{O \cap A^c \mid O \in \mathcal{O},\ A \text{ is countable}\}.$$

Then $\mathcal{O}_\tau$ satisfies the axioms of open sets as checked below: $\mathcal{O}_\tau$ is closed under finite intersection, because a finite union of countable sets is countable. $\mathcal{O}_\tau$ is closed under arbitrary union, because: For every family $\{O_\lambda\}_{\lambda\in\Lambda} \subset \mathcal{O}$, there exists a countable subfamily $\{O_{\lambda_n}\}$ such that $\bigcup_{\lambda\in\Lambda} O_\lambda = \bigcup_{n=1}^{\infty} O_{\lambda_n}$ (Lindelöf property). Therefore $\bigcup_{\lambda\in\Lambda}(O_\lambda \cap A_\lambda^c)$ is contained in $\bigcup_{\lambda\in\Lambda} O_\lambda = \bigcup_{n=1}^{\infty} O_{\lambda_n}$, and contains $\bigcup_{n=1}^{\infty}(O_{\lambda_n} \cap A_{\lambda_n}^c)$ hence contains $\bigcup_{n=1}^{\infty} O_{\lambda_n} \cap (\bigcup_{n=1}^{\infty} A_{\lambda_n})^c$. Thus we get $\bigcup_{\lambda\in\Lambda}(O_\lambda \cap A_\lambda^c) \in \mathcal{O}_\tau$.

Consider a topology τ in which $\mathcal{O}_\tau$ is the family of all open sets. Since $\mathcal{O} \subset \mathcal{O}_\tau \subset \mathcal{B}$ (=the usual Borel field of Euclidean topology), we have $\mathcal{B}(\mathcal{O}_\tau) = \mathcal{B}$. Namely the Borel field of τ is identical with $\mathcal{B}$.

Though $(\mathbb{R}^n,\mathcal{B})$ is standard, $(\mathbb{R}^n,\tau)$ is not Luzin. Since a Luzin space is continuous image of a separable space, it must be separable. But $(\mathbb{R}^n,\tau)$ is not separable, because every countable set is closed in τ.

Besides the case of a Borel measurable space, we have no

useful criterion for the standardness of a general measurable space. We can easily see that a standard measurable space $(X, \mathcal{B})$ is σ-separated, but the converse is not true. A counter example is given by $(Y, \mathcal{B} \cap Y)$, Y being a non-Borel set of $\mathbb{R}$ and $\mathcal{B}$ being the Borel field of $\mathbb{R}$. However:

Remark. If $(X, \mathcal{B})$ is σ-separated and if there exists a measurable bijection f from a standard measurable space, then f^{-1} is also measurable as shown below, so that $(X, \mathcal{B})$ is standard. Suppose that $\{B_n\} \subset \mathcal{B}$ separates X, then the map Φ: $X \ni x \to (c_n(x)) \in \{0,1\}^{\infty} = M$ is measurable. Then $g = \Phi \circ f$ is a measurable injection from a standard space into $(M, \mathcal{B}_M)$, so that g^{-1} is measurable by Th. 14.3 b). Thus $f^{-1} = g^{-1} \circ \Phi$ is measurable.

CHAPTER 3. MEASURES ON VECTOR SPACES

§15. Explanation of the problem

We want to construct a measure on a real vector space X.

For linear functions $\xi_1,\xi_2,\cdots,\xi_n$ on X, suppose that the distribution of $(\xi_1(x),\xi_2(x),\cdots,\xi_n(x))$ is given in a self-consistent way. The self-consistency conditions should be as follows:

First, $(\xi_1(x),\xi_2(x),\cdots,\xi_n(x))\in B$ is equivalent to $(\xi_1(x),\xi_2(x),\cdots,\xi_n(x),\xi_{n+1}(x))\in B\times\mathbf{R}$, therefore we should have

$$(15.1)\qquad \mu_{\xi_1\xi_2\cdots\xi_n}(B) = \mu_{\xi_1\xi_2\cdots\xi_n\xi_{n+1}}(B\times\mathbf{R})$$

$$\text{for } {}^\forall B:\ \text{Borel set of } \mathbf{R}^n.$$

Second, let $T=(t_{ij})$ be a real non-singular matrix. If $\xi_i=\sum_{j=1}^n t_{ij}\eta_j$, then $(\xi_1(x),\xi_2(x),\cdots,\xi_n(x))\in B$ is equivalent to $(\eta_1(x),\eta_2(x),\cdots,\eta_n(x))\in T^{-1}(B)$, therefore we should have

$$(15.2)\qquad \xi_i = \sum_{j=1}^n t_{ij}\eta_j \Rightarrow \mu_{\xi_1\xi_2\cdots\xi_n}(B) = \mu_{\eta_1\eta_2\cdots\eta_n}(T^{-1}(B)).$$

Under two assumptions (15.1) and (15.2), our problem is to construct a measure on X such that

$$(15.3)\qquad \mu_{\xi_1\xi_2\cdots\xi_n}(B) = \mu(\{x\in X\,|\,(\xi_1(x),\xi_2(x),\cdots,\xi_n(x))\in B\})$$

$${}^\forall\xi_1,\xi_2,\cdots,\xi_n \text{ and } {}^\forall B:\ \text{Borel set of } \mathbf{R}^n.$$

For a rigorous formulation, we need two more requirements. First, we must decide the limits of the choice of $\xi_1,\xi_2,\cdots,\xi_n$. Let X^a be the algebraical dual space of X, and E be any fixed subspace of X^a. We shall assume that $\xi_1,\xi_2,\cdots,\xi_n$

should be chosen in E. Next, we must decide the σ-algebra on which μ is defined. In order that (15.3) makes sense, the set $\{x \in X \mid (\xi_1(x), \xi_2(x), \cdots, \xi_n(x)) \in B\}$ should be measurable. In other words, the function $\xi(x)$ should be measurable for every $\xi \in E$. Let $\mathcal{B}_E$ be the smallest σ-algebra on X in which all $\xi(x)$ $(\xi \in E)$ are measurable,

$$(15.4) \qquad \mathcal{B}_E = \mathcal{B}\Big(\bigcup_{\xi \in E} \xi^{-1}(\mathcal{B}_0)\Big)$$

where $\mathcal{B}_0$ is the Borel field of $\mathbf{R}$. We want to construct the measure on $\mathcal{B}_E$.

Now, our problem has been completely prescribed. In general, the answer of this problem is negative. We need some additional conditions on $\{\mu_{\xi_1 \xi_2 \cdots \xi_n}\}$ to extend it to a σ-additive measure on $(X, \mathcal{B}_E)$. The purpose of this chapter is to discuss such conditions on $\{\mu_{\xi_1 \xi_2 \cdots \xi_n}\}$.

We shall rewrite our problem in another way to apply the results of the previous chapter. First, we shall imbed X into E^a, the algebraical dual space of E. Since $\xi(x)$ is bilinear with respect to ξ and x, every $x \in X$ induces a linear map on E. This linear map is zero if and only if $x \in E^{\perp} = \{x \in X \mid \xi(x) = 0 \text{ for } {}^\forall \xi \in E\}$. Thus, the factor space $X/E^{\perp}$ can be imbedded into E^a. As we shall explain below, we can assume $E^{\perp} = \{0\}$. In this case, X can be imbedded into E^a.

In general, every $\xi \in E$ becomes a linear function on $X/E^{\perp}$. Namely, denoting the canonical surjection from X onto $X/E^{\perp}$ with p, we have ${}^\exists \bar{\xi} \in (X/E^{\perp})^a, \xi = \bar{\xi} \circ p$. Thus, from the definition of $\mathcal{B}_E$, we have $\mathcal{B}_E = p^{-1}(\bar{\mathcal{B}}_E)$ where $\bar{\mathcal{B}}_E = \mathcal{B}\big(\bigcup_{\xi \in E} \bar{\xi}^{-1}(\mathcal{B}_0)\big)$. On the other hand, in (15.3) we have

$$\{x\in X\,|\,(\xi_1(x),\xi_2(x),\cdots,\xi_n(x))\in B\}$$
$$= p^{-1}(\{\bar{x}\in X/E^{\perp}\,|\,(\bar{\xi}_1(\bar{x}),\bar{\xi}_2(\bar{x}),\cdots,\bar{\xi}_n(\bar{x}))\in B\}).$$

Thus, by the one-to-one correspondence $\mu\leftrightarrow p\circ\mu$ between measures on $(X,\mathcal{B}_E)$ and those on $(X/E^{\perp},\bar{\mathcal{B}}_E)$, our extension problem for X is reduced to the same problem for $X/E^{\perp}$. So, replacing X by $X/E^{\perp}$ if necessary, from the first we can assume $E^{\perp}=\{0\}$. Throughout this chapter, we shall always assume this.

Now, X is regarded as a subspace of E^a. Then we have:

Theorem 15.1. A self-consistent family $\{\mu_{\xi_1\xi_2\cdots\xi_n}\}$ (this means that $\{\mu_{\xi_1\xi_2\cdots\xi_n}\}$ satisfies (15.1) and (15.2)) corresponds in a one-to-one way to a σ-additive measure μ on $(E^a, \mathcal{B}'_E)$. Here, the correspondence is given by

$$(15.5)\qquad \mu_{\xi_1\xi_2\cdots\xi_n}(B) = \mu(\{f\in E^a\,|\,(f(\xi_1),f(\xi_2),\cdots,f(\xi_n))\in B\}),$$

and $\mathcal{B}'_E$ is the smallest σ-algebra in which $f\to f(\xi)$ is measurable for every $\xi\in E$. (Hence, evidently we have $\mathcal{B}_E=\mathcal{B}'_E\cap X$).

Proof. It is evident that for any measure μ on $(E^a,\mathcal{B}'_E)$, the corresponding $\{\mu_{\xi_1\xi_2\cdots\xi_n}\}$ satisfies (15.1) and (15.2). We shall prove the converse.

Let R be a finite dimensional subspace of E, and R^a be its algebraical dual space. The condition (15.2) means that we can consider a measure μ_R on R^a independently of the choice of its base. More detailed discussions are as follows: Suppose that $\xi_1,\xi_2,\cdots,\xi_n$ is a base of R, and put

$$(15.6)\qquad \mu_R(\{f\in R^a\,|\,(f(\xi_1),f(\xi_2),\cdots,f(\xi_n))\in B\}) = \mu_{\xi_1\xi_2\cdots\xi_n}(B).$$

For a real non-singular matrix $T=(t_{ij})$, if $\xi_i=\sum_{j=1}^{n} t_{ij}\eta_j$, then

$(f(\xi_1),f(\xi_2),\cdots,f(\xi_n))\in B$ is equivalent to $(f(\eta_1),f(\eta_2),\cdots,$ $f(\eta_n))\in T^{-1}(B)$. Therefore the condition (15.2) means that μ_R is independent of the choice of $\xi_1,\xi_2,\cdots,\xi_n$.

For $R'\subset R$ and $f\in R^a$, let $\pi_{R'R}f$ be the restriction of f on R'. Evidently $\pi_{R'R}$ is linear, surjective and Borel measurable, because on a finite dimensional vector space every linear map is Borel measurable. For $R''\subset R'\subset R$, we have evidently $\pi_{R''R}=\pi_{R''R'}\circ\pi_{R'R}$. Thus $\{(R^a,\mathcal{B}_R),\pi_{R'R}\}$ becomes a projective family of measurable spaces in the sense of §8. ($\mathcal{B}_R$ is the Borel field of R^a).

The condition (15.1) means that $\{\mu_R\}$ is a self-consistent family, because if we choose a base $\xi_1,\xi_2,\cdots,\xi_n$ of R such that $\xi_1,\xi_2,\cdots,\xi_m$ $(m<n)$ belong to R', then $(\pi_{R'R}f(\xi_1),$ $\pi_{R'R}f(\xi_2),\cdots,\pi_{R'R}f(\xi_m))\in B$ is equivalent to $(f(\xi_1),f(\xi_2),\cdots,$ $f(\xi_n))\in B\times\mathbb{R}^{n-m}$, thus (15.1) implies $\mu_{R'}=\pi_{R'R}\circ\mu_R$.

For $f\in E^a$, let $\pi_R f$ be the restriction of f on R. Then, (15.5) is equivalent to $\mu_R=\pi_R\circ\mu$. So, our problem (15.5) has been reduced to the extension problem of $\{\mu_R\}$.

Since each $R^a\cong\mathbb{R}^n$ is a σ-compact metric space, by the corollary of Th.10.1, $\{\mu_R\}$ is extended uniquely to a σ-additive measure on the projective limit measurable space. Therefore the proof will be completed if we show that $(E^a,\mathcal{B}_E')$ is isomorphic to the projective limit measurable space of $(R^a,\mathcal{B}_R)$.

For a given $f\in E^a$, $(\pi_R f)\in\prod_R R^a$ belongs to the projective limit, because we have evidently $\pi_{R'R}(\pi_R f)=\pi_{R'}f$. Conversely if $(f_R)\in\prod_R R^a$ satisfies $\pi_{R'R}f_R=f_{R'}$, then defining f by $f(\xi)=f_R(\xi)$ for $\xi\in R$, f becomes a linear map on E, namely

$f \in E^a$ and $f_R = \pi_R f$. This means that the projective limit is identified with E^a as a set. The projective limit σ-algebra is the smallest one in which $(f_R) \to f_R$ is measurable, while $\mathcal{B}'_E$ is the smallest one in which $f \to \pi_R f$ is measurable. This means that the one-to-one correspondence $f \leftrightarrow (\pi_R f)$ is a measurable isomorphism between $(E^a, \mathcal{B}'_E)$ and the projective limit measurable space of $(R^a, \mathcal{B}_R)$. (q.e.d.)

<u>Remark</u>. Consider a sequence $R_1 \subset R_2 \subset \cdots \subset R_n \subset \cdots$. Putting $M = \bigcup_{n=1}^{\infty} R_n$, the projective limit of $(R_n^a, \mathcal{B}_{R_n})$ is isomorphic to $(M^a, \mathcal{B}_M)$. Since every linear function on M can be extended to a linear function on E, the map π_M is surjective from E^a onto M^a. Thus the corollary of Th.10.1 can be applied.

<u>Theorem 15.2</u>. A self-consistent family $\{\mu_{\xi_1 \xi_2 \cdots \xi_n}\}$ can be extended to a measure on $(X, \mathcal{B}_E)$, if and only if X is thick in E^a with respect to the measure μ defined by (15.5).

<u>Proof</u>. If X is thick with respect to μ, then consider the trace ν of μ on X. Denoting the imbedding $X \to E^a$ with i, we have $\mu = i \circ \nu$, so that

$$\mu_{\xi_1 \xi_2 \cdots \xi_n}(B) = \mu(\{f \in E^a \mid (f(\xi_1), f(\xi_2), \cdots, f(\xi_n)) \in B\})$$

$$= \nu(\{x \in X \mid (\xi_1(x), \xi_2(x), \cdots, \xi_n(x)) \in B\}).$$

Thus $\{\mu_{\xi_1 \xi_2 \cdots \xi_n}\}$ is extended to the measure ν on $(X, \mathcal{B}_E)$.

Conversely, suppose that $\{\mu_{\xi_1 \xi_2 \cdots \xi_n}\}$ is extended to some measure ν on $(X, \mathcal{B}_E)$. Imbed X into E^a and put $\mu = i \circ \nu$. Then by (15.3) and (15.6) we have

$$\mu_R(\{f \in R^a \mid (f(\xi_1), f(\xi_2), \cdots, f(\xi_n)) \in B\})$$

$$= \mu(\{f \in E^a \mid (f(\xi_1), f(\xi_2), \cdots, f(\xi_n)) \in B\}).$$

This means $\mu_R = \pi_R \circ \mu$. From the uniqueness of the extension, the measure μ is identical with the one given in (15.5). Since $\mu = i \circ \nu$, X is thick in μ and ν is the trace of μ on X. (q.e.d.)

Remark. If E is countable dimensional, then every one-point set of E^a is $\mathcal{B}'_E$-measurable. So, unless $X = E^a$, some $\{\mu_{\xi_1 \xi_2 \cdots \xi_n}\}$ can not be extended to a measure on $(X, \mathcal{B}_E)$. In the case that E is non-countable dimensional, if and only if $\pi_M(X) = M^a$ for every countable dimensional $M \subset E$, every self-consistent $\{\mu_{\xi_1 \xi_2 \cdots \xi_n}\}$ is extended to a measure on $(X, \mathcal{B}_E)$. This comes from $\mathcal{B}_E = \bigcup_M \pi_M^{-1}(\mathcal{B}_M)$ as seen in §8.

By Th.15.2, our extension problem becomes equivalent to the evaluation of the carrier of the measure μ.

§16. Relation with Bochner's theorem

For finite dimensional case, the classical Bochner's theorem assures a one-to-one correspondence between measures and characteristic functions. First, we shall explain the classical Bochner's theorem without proof.

Let μ be a Borel measure on $\mathbb{R}^n$. The characteristic function of μ is defined by

$$\chi(\xi) = \int_{\mathbb{R}^n} \exp(i(\xi, x))\, d\mu(x), \tag{16.1}$$

where (ξ, x) is the usual inner product on $\mathbb{R}^n$. In other words, $\chi(\xi)$ is the Fourier transform of μ.

Evidently $\chi(\xi)$ is a continuous function. It is positive definite, namely it satisfies:

(16.2) $\forall \xi_1, \xi_2, \cdots, \xi_N$ and $\forall \alpha_1, \alpha_2, \cdots, \alpha_N$: complex numbers

$$\sum_{j,k=1}^{N} \alpha_j \bar{\alpha}_k \chi(\xi_j - \xi_k) \geqq 0.$$

(16.2) is true since the left side of (16.2) is equal to

$$\int_{\mathbb{R}^n} \left| \sum_{j=1}^{N} \alpha_j \exp(i(\xi_j, x)) \right|^2 d\mu(x).$$

The converse is the classical Bochner's theorem.

Theorem 16.1. For every continuous and positive definite function $\chi(\xi)$ on $\mathbb{R}^n$, there exists uniquely a Borel measure μ on $\mathbb{R}^n$ which satisfies (16.1).

Proof is omitted here. (For instance c.f. S.Bochner; "Lectures on Fourier integrals" Princeton, 1959).

Now, we shall define and discuss characteristic functions for infinite dimensional case.

Definition 16.1. Let E be a real vector space. A function $\chi(\xi)$ on E is called a characteristic function, if it is positive definite (namely satisfies (16.2)) and continuous on every finite dimensional $R \subset E$ (with respect to the Euclidean topology on R).

Theorem 16.2. A characteristic function $\chi(\xi)$ on E corresponds in a one-to-one way with a measure μ on $(E^a, \mathcal{B}'_E)$ by the following correspondence:

(16.3) $$\chi(\xi) = \int_{E^a} \exp(if(\xi))\, d\mu(f).$$

(E^a is the algebraic dual space of E, and $\mathcal{B}'_E$ is the smallest σ-algebra in which $f \to f(\xi)$ is measurable for every $\xi \in E$).

Proof. For a given measure μ on $(E^a, \mathcal{B}'_E)$, the corresponding $\chi(\xi)$ is positive definite as shown below (16.2). For a finite dimensional $R \subset E$, (16.3) implies

$$\chi(\xi) = \int_{R^a} \exp(if(\xi))\, d(\pi_R \circ \mu)(f) \quad \text{for } \forall \xi \in R.$$

Thus the restriction of $\chi(\xi)$ on R is the characteristic function of the finite dimensional measure $\pi_R \circ \mu$, therefore it is continuous on R.

Conversely, suppose that $\chi(\xi)$ is positive definite and continuous on every R. Since the restriction of $\chi(\xi)$ on R is continuous and positive definite, by the classical Bochner's theorem, there exists uniquely a measure μ_R on R^a such that

$$(16.4) \qquad \chi(\xi) = \int_{R^a} \exp(if(\xi))\, d\mu_R(f) \quad \text{for } \forall \xi \in R.$$

The family $\{\mu_R\}$ is self-consistent as shown below: If $R \subset R_1$, we have for $\xi \in R$

$$\chi(\xi) = \int_{R^a} \exp(if(\xi))\, d\mu_R(f)$$

$$= \int_{R_1^a} \exp(if(\xi))\, d\mu_{R_1}(f) = \int_{R^a} \exp(if(\xi))\, d(\pi_{RR_1} \circ \mu_{R_1})(f).$$

By the one-to-one correspondence between χ and μ for the finite dimensional case, we have $\mu_R = \pi_{RR_1} \circ \mu_{R_1}$. Thus $\{\mu_R\}$ is a self-consistent family.

Therefore $\{\mu_R\}$ can be extended uniquely to a σ-additive measure μ on $(E^a, \mathcal{B}'_E)$, i.e. we have $\mu_R = \pi_R \circ \mu$. It is evident that (16.3) is satisfied for this μ (and only for this μ).

(q.e.d.)

Let τ be a vector topology on E. (A vector topology is a topology in which E becomes a topological vector space).

A generalization of the classical Bochner's theorem would be: If $\chi(\xi)$ is continuous in τ, then the corresponding measure μ is constructed on the topological dual space E'. But this "theorem" is false in general.

For instance, even if $\chi(\xi)$ is continuous on a Hilbert space H, the corresponding measure may not lie on its dual space H', but it lies on a Hilbert-Schmidt extension of H'. If $\chi(\xi)$ is continuous on a nuclear space E, then the generalized Bochner's theorem holds, and the corresponding measure lies on E'.

Anyway, the relation between the continuity of χ and the carrier of μ is an important topic, which we shall study in this chapter.

Before closing this section, we shall mention some properties of positive definite functions. $\chi(\xi)$ is positive definite if and only if the matrix $(\chi(\xi_j-\xi_k))_{1\leqq j,k\leqq n}$ is positive definite for every n and every $\xi_1,\xi_2,\cdots,\xi_n$. Take $n=1$, then we see that $\chi(0)\geqq 0$. Take $n=2$, then we see that

(16.5) $\quad \chi(-\xi) = \overline{\chi(\xi)}$ and $|\chi(\xi)| \leqq \chi(0)$.

Take $n=3$, then we see that

(16.6) $\quad |\chi(\xi + \eta) - \chi(\xi)| \leqq \sqrt{2\chi(0)}\sqrt{|\chi(\eta) - \chi(0)|}$.

From (16.6), we see that if $\chi(\xi)$ is continuous at the origin in a vector topology τ, then $\chi(\xi)$ is uniformly continuous in τ.

§17. Minlos' theorem

In this section, we shall discuss Bochner's theorem for a Hilbert space. We shall start discussions with a counter example.

Example 17.1. Let H be a Hilbert space. The function $\chi(\xi) = \exp(-\|\xi\|^2/2)$ is positive definite, because on every finite dimensional $R \subset H$, $\chi(\xi)$ becomes the characteristic function of a (finite dimensional) gaussian measure. So by Th.16.2, $\chi(\xi)$ is the characteristic function of a measure μ on $(H^a, \mathcal{B}_H)$, which we shall call the infinite dimensional gaussian measure.

Let H' be the topological dual space of H. For an orthonormal system (=ONS) $\{e_n\}$ of H, the distribution of $f(e_n)$ becomes an i.i.d. (=independent identical distribution, in this case the one-dimensional gaussian measure g_1). So that we have $\mu(\{f \in H^a \mid \sup_n |f(e_n)| \leq M\}) = g_1([-M,M])^\infty = 0$. Letting $M \to \infty$, we see that $\mu(\{f \in H^a \mid \sup_n |f(e_n)| < \infty\}) = 0$, namely $\{f(e_n)\}$ is not bounded for almost all f. But if $f \in H'$, we have $\{f(e_n)\} \in (\ell^2)$, thus H' is a null set with respect to μ.

Definition 17.1. Let H be a Hilbert space, and T be a bounded operator on H. T is called a Hilbert-Schmidt (=HS) operator, if it satisfies:

$$\sum_{\lambda} \|Te_\lambda\|^2 < \infty, \tag{17.1}$$

where $\{e_\lambda\}$ is a complete orthonormal system (=CONS) of H.

We shall remark that the left side of (17.1) is independent of the choice of $\{e_\lambda\}$. For another CONS $\{e'_\mu\}$, we have $\sum_\lambda \|Te_\lambda\|^2 = \sum_\lambda\sum_\mu (Te_\lambda, e'_\mu)^2 = \sum_\mu\sum_\lambda (e_\lambda, T^*e'_\mu)^2 = \sum_\mu \|T^*e'_\mu\|^2$, therefore denoting the left side of (17.1) with $|||T|||^2$, we see that $|||T||| = |||T^*|||$ and that it does not depend on $\{e_\lambda\}$.

Theorem 17.1 (Minlos). Let H be a Hilbert space, and T be an HS operator. Put $\|\xi\|_1 = \|T\xi\|$, and suppose that $\chi(\xi)$ is positive definite and continuous in $\|\xi\|_1$. Then, the corresponding measure lies on H'.

Proof. We shall prove that H' is thick in H^a. For this purpose, it is sufficient to show that for every countable dimensional $M \subset H$, the set $\pi_M(H')$ is thick in M^a with respect to $\pi_M \circ \mu$, or equivalently $B = \pi_M^{-1}(\pi_M(H'))$ is thick in H^a. $f \in B$ means that $\exists g \in H'$, $\pi_M f = \pi_M g$, namely means that the restriction of f on M is continuous. Therefore for an orthonormal base $\{e_n\}$ of M, we have

$$(17.2) \qquad B = \{f \in H^a \mid \sum_{n=1}^{\infty} f(e_n)^2 < \infty\}.$$

Since we have known $B \in \mathcal{B}_H$, it is thick if and only if $\mu(B) = \mu(H^a)$ $(= \chi(0))$. We shall prove this equality. From (17.2) we get

$$(17.3) \qquad \mu(B) = \lim_{\alpha \to +0} \lim_{N \to \infty} \int_{H^a} \exp(-\frac{\alpha}{2} \sum_{n=1}^{N} f(e_n)^2)\, d\mu(f),$$

because $\lim_{\alpha \to +0} \lim_{N \to \infty} \exp(-\frac{\alpha}{2} \sum_{n=1}^{N} f(e_n)^2) = 1$ for $f \in B$, and $= 0$ for $f \notin B$, and the integrand is less than the constant 1, allowing us to apply Lebesgue's convergence theorem.

We shall denote the integral in (17.3) with $I_{\alpha,N}$. Denoting the distribution of $(f(e_1), f(e_2), \cdots, f(e_N))$ with μ_N, we have

$$I_{\alpha,N} = \int_{\mathbb{R}^N} \exp(-\frac{\alpha}{2} \sum_{n=1}^{N} t_n^2)\, d\mu_N(t).$$

Taking the Fourier transform, we can rewrite it as:

$$I_{\alpha,N} = (2\pi\alpha)^{-N/2}\int_{\mathbb{R}^N} \exp\left(-\frac{1}{2\alpha}\sum_{n=1}^{N} s_n^2\right)\chi_N(s)\, ds_1 ds_2 \cdots ds_N,$$

where $\chi_N(s)$ is the characteristic function of μ_N, and is calculated as:

$$\begin{aligned}\chi_N(s_1, s_2, \cdots, s_N) &= \int_{\mathbb{R}^N} \exp\left(i\sum_{n=1}^{N} s_n t_n\right) d\mu_N(t) \\ &= \int_{H^a} \exp\left(i\sum_{n=1}^{N} s_n f(e_n)\right) d\mu(f) \\ &= \int_{H^a} \exp\left(if\left(\sum_{n=1}^{N} s_n e_n\right)\right) d\mu(f) = \chi\left(\sum_{n=1}^{N} s_n e_n\right).\end{aligned}$$

Thus, we get

$$(17.4) \quad I_{\alpha,N} = (2\pi\alpha)^{-N/2}\int_{\mathbb{R}^N} \exp\left(-\frac{1}{2\alpha}\sum_{n=1}^{N} s_n^2\right)\chi\left(\sum_{n=1}^{N} s_n e_n\right) ds_1 ds_2 \cdots ds_N.$$

On the other hand, since $\chi(\xi)$ is continuous in $\|\xi\|_1$, for every $\varepsilon>0$ there exists $\delta>0$ such that $\|\xi\|_1<\delta$ implies $|\chi(0)-\chi(\xi)|<\varepsilon$. Combining this with $|\chi(\xi)|\leqq\chi(0)$, we have $\mathcal{Re}\,\chi(\xi)>\chi(0)-\varepsilon-2\chi(0)\frac{\|\xi\|_1^2}{\delta^2}$ for $\forall\xi\in H$. Substituting this into (17.4), we get

$$(17.5) \quad I_{\alpha,N} > \chi(0) - \varepsilon - 2\chi(0)\frac{1}{\delta^2}E\left(\left\|\sum_{n=1}^{N} s_n e_n\right\|_1^2\right),$$

where $E(\ )$ is the expectation with respect to the independent gaussian distribution with variance α. Since $\|\sum_{n=1}^{N} s_n e_n\|_1^2 = \|\sum_{n=1}^{N} s_n Te_n\|^2 = \sum_{n=1}^{N}\sum_{m=1}^{N} s_n s_m (Te_n, Te_m)$, and since $E(s_n s_m)=\alpha\delta_{nm}$, we have $E(\|\sum_{n=1}^{N} s_n e_n\|_1^2)=\alpha\sum_{n=1}^{N}\|Te_n\|^2$, hence

$$(17.6) \quad I_{\alpha,N} > \chi(0) - \varepsilon - 2\chi(0)\frac{\alpha}{\delta^2}\sum_{n=1}^{N}\|Te_n\|^2.$$

Under the fixed ε and δ, take the limit $N\to\infty$ and then $\alpha\to+0$. Since $\sum_{n=1}^{\infty}\|Te_n\|^2<\infty$ from the assumption, we see that

(17.7) $$\mu(B) = \lim_{\alpha\to+0}\lim_{N\to\infty} I_{\alpha,N} \geqq \chi(0) - \varepsilon.$$

Since $\varepsilon>0$ is arbitrary, this assures $\mu(B)=\chi(0)$, thus the proof has been completed. (q.e.d.)

We can rewrite Th.17.1 without using the operator T.

Let $(\, , \,)$ and $(\, , \,)_1$ be two inner products on a real vector space E. (An inner product is defined as a symmetric bilinear function on $E\times E$ which satisfies $(\xi,\xi)\geqq 0$ for $\forall\xi\in E$).

<u>Definition 17.2</u>. $(\, , \,)_1$ is said to be HS (=Hilbert-Schmidt) with respect to $(\, , \,)$, if there exists $M>0$ such that for every finite ONS $\{\xi_n\}$ in $(\, , \,)$, we have

(17.8) $$\sum_n \|\xi_n\|_1^2 \leqq M^2.$$

The HS property of an inner product has the following relation with an HS operator. If E is a Hilbert space with respect to $(\, , \,)$, and if $(\xi,\eta)_1=(T\xi,T\eta)$ for some HS operator T, then the inner product $(\, , \,)_1$ is HS with respect to $(\, , \,)$.

Conversely, suppose that $(\, , \,)_1$ is HS with respect to $(\, , \,)$. Then, especially we have $\|\xi\|_1\leqq M\|\xi\|$. Put $F=\{\xi\in E \mid \|\xi\|=0\}$, then the factor space E/F becomes a pre-Hilbert space with respect to $(\, , \,)$. Let H be the completion of E/F. The inner product $(\, , \,)_1$ is also regarded as an inner product on E/F, and since $|(\xi,\eta)_1|\leqq\|\xi\|_1\|\eta\|_1\leqq M^2\|\xi\|\,\|\eta\|$, there exists a bounded operator S on H such that $(\xi,\eta)_1=(\xi,S\eta)$. Since S is a non-negative symmetric operator, we can consider

$\sqrt{S}=T$. (Consider the spectral resolution $S=\int_{[0,\infty)}\lambda dP_\lambda$ and put $T=\int_{[0,\infty)}\sqrt{\lambda}dP_\lambda$). Then we have $(\xi,\eta)_1=(T\xi,T\eta)$. Taking a CONS of H in E/F, (17.8) implies that T is an HS operator on H. Thus we get: $\|\xi\|_1=\|T\xi\|$, where T is an HS operator on H (=the completion of E/F with respect to $\|\cdot\|$).

Theorem 17.2. Let E be a vector space, E_0 be a subspace of E (E_0 may be equal to E), and $(\, , \,)_1$ (resp. $(\, , \,)_0$) be an inner product on E (resp. E_0). Suppose that a positive definite function $\chi(\xi)$ on E is continuous in $\|\cdot\|_1$, and suppose that $(\, , \,)_1$ is HS on E_0 with respect to $(\, , \,)_0$. Then the corresponding measure lies on E_0', where

(17.9)

$$E_0' = \{f\in E^a \mid \text{restriction of } f \text{ on } E_0 \text{ is continuous in } \|\cdot\|_0\}.$$

Proof is quite the same as that of Th.17.1. Only required modifications are: (1) to use $\|\xi\|_1$ instead of $\|T\xi\|$ and $(\xi,\eta)_1$ instead of $(T\xi,T\eta)$, (2) to replace H^a by E^a and H' by E_0', and (3) to suppose that $\{e_n\}$ is an orthonormal base of $M\cap E_0$ (in $(\, , \,)_0$).

We shall apply Th.17.2 to a Hilbert space H with the norm $\|\cdot\|$. If we take $\|\xi\|_0=\|\xi\|$ and $\|\xi\|_1=\|T\xi\|$, then we have Th.17.1. But if we take $\|\xi\|_1=\|\xi\|$, then we have another theorem.

Theorem 17.3. Let H be a Hilbert space, and suppose that $\chi(\xi)$ is continuous on H. Let H_0 be a subspace of H (H_0 may be equal to H), and $(\, , \,)_0$ be an inner product on H_0 such that $(\, , \,)$ is HS on H_0 with respect to $(\, , \,)_0$. Then the corresponding measure lies on H_0', where H_0' is given by (17.9) replacing E by H.

Example 17.2. Take $E=\mathbb{R}_0^\infty$, then $E^a=\mathbb{R}^\infty$. Suppose that $\chi(\xi)$ is continuous in (ℓ^2)-norm, then the corresponding measure may not lie on (ℓ^2), but it lies on $(\ell^2)_a=\{x\in\mathbb{R}^\infty \mid \sum_{n=1}^{\infty} a_n^2x_n^2<\infty\}$ for every $a=(a_n)\in(\ell^2)$. Because the (ℓ^2)-norm is HS on $\mathbb{R}_0^\infty$ with respect to $(\xi,\eta)_0=\sum_{n=1}^{\infty}\frac{\xi_n\eta_n}{a_n^2}$, and $(\ell^2)_a$ is the topological dual space of $\mathbb{R}_0^\infty$ in the norm $\|\cdot\|_0$.

§18. Sazonov topology

First, we shall prove the converse of Th.17.1.

Definition 18.1. A semi-norm $\|\cdot\|$ on E is said to be Hilbertian if it is induced by some inner product $(\ ,\ )$, namely if $\|\xi\|^2=(\xi,\xi)$ for $^\forall\xi\in E$. Let $\|\cdot\|$ and $\|\cdot\|_1$ be two Hilbertian semi-norms. $\|\cdot\|_1$ is said to be HS with respect to $\|\cdot\|$, if the corresponding $(\ ,\ )_1$ is HS with respect to $(\ ,\ )$.

Definition 18.2. Let $(X,\mathcal{B},\mu)$ be a measure space, and U be a subset of X. A $\mathcal{B}$-measurable set $B_0\supset U$ is called an equi-measure cover of U, if it satisfies $\mu^*(U)=\mu(B_0)$. (μ^* is the outer measure of μ).

An equi-measure cover always exists from the σ-additivity of μ.

Theorem 18.1 (Sazonov). Let H be a Hilbert space. Suppose that μ lies on H', then the characteristic function of μ is continuous in some Hilbertian semi-norm $\|\cdot\|_1$ which is HS with respect to $\|\cdot\|$.

Proof. Let U be the unit ball of H', then we have $H'=\bigcup_{n=1}^{\infty}(nU)$. Let B_n be an equi-measure cover of nU, then we have $\lim_{n\to\infty}\mu(B_n^c)=0$. Put

$$(18.1)\qquad (\xi,\eta)_1 = \sum_{n=1}^{\infty} \frac{1}{n^4}\int_{B_n} f(\xi)f(\eta)\, d\mu(f).$$

For every finite ONS $\{\xi_k\}$ of H, we have $\sum_k f(\xi_k)^2 \leqq \|f\|'^2$. ($\|\cdot\|'$ is the dual norm of $\|\cdot\|$). So that for almost all $f \in B_n$, we have $\sum_k f(\xi_k)^2 \leqq n^2$. Therefore we get

$$\sum_k \|\xi_k\|_1^2 = \sum_{n=1}^{\infty} \frac{1}{n^4}\int_{B_n} \sum_k f(\xi_k)^2\, d\mu(f) \leqq \sum_{n=1}^{\infty} \frac{1}{n^2}\mu(H^a) \equiv M < \infty.$$

Thus, $\|\cdot\|_1$ is HS with respect to $\|\cdot\|$.

On the other hand, from the definition of χ, we have

$$\chi(\xi) = \int_{H^a} \exp(if(\xi))\, d\mu(f),$$

so that we have

$$(18.2)\qquad |\chi(0) - \chi(\xi)| \leqq \int_{B_n} |1 - \exp(if(\xi))|\, d\mu(f) + 2\mu(B_n^c).$$

For a given $\varepsilon>0$, we shall choose n such that $\mu(B_n^c)<\varepsilon$. For this n, $\|\xi\|_1^2 \leqq \frac{\varepsilon^2}{n^4}$ implies $\int_{B_n} f(\xi)^2\, d\mu(f) \leqq \varepsilon^2$. Since $|1-\exp(if(\xi))| \leqq |f(\xi)|$, applying Schwartz's inequality, we get from (18.2)

$$|\chi(0) - \chi(\xi)| \leqq \sqrt{\mu(B_n)}\varepsilon + 2\varepsilon.$$

Thus we see that $\|\xi\|_1 \leqq \frac{\varepsilon}{n(\varepsilon)^2}$ implies $|\chi(0)-\chi(\xi)| \leqq (2+\sqrt{\mu(H^a)})\varepsilon$. This means that χ is continuous in $\|\cdot\|_1$. (q.e.d.)

The theorems 17.2 and 18.1 can be generalized to the case of a family of Hilbertian semi-norms.

<u>Definition 18.3</u>. A topological vector space E is called a Hilbertian type space, if its topology is defined by a family

of Hilbertian semi-norms. (We shall always assume that the family of semi-norms is directed, namely if $\|\cdot\|_1$ and $\|\cdot\|_2$ belong to this family, then some $\|\cdot\|_3 \geq \|\cdot\|_1, \|\cdot\|_2$ belongs to this family). Let σ and τ be two Hilbertian type topologies on E. σ is said to be HS with respect to τ, if every inner product in σ is HS with respect to some inner product in τ.

The definition of HS property for Hilbertian type topologies does not depend on the choice of the family of inner products defining σ or τ.

Theorem 18.2. Let E be a vector space, E_0 be a subspace of E, and σ (resp. τ) be a Hilbertian type topology on E (resp. E_0). Suppose that a positive definite function $\chi(\xi)$ on E is continuous in σ, and suppose that σ is HS on E_0 with respect to τ. Then the corresponding measure lies on E'_τ, where

(18.3)
$$E'_\tau = \{f \in E^a \mid \text{restriction of } f \text{ on } E_0 \text{ is continuous in } \tau\}.$$

Proof is quite similar to that of Th.17.1. Suppose that σ is defined by $\{\|\cdot\|_{\sigma\alpha}\}_{\alpha \in A}$, and τ is defined by $\{\|\cdot\|_{\tau\beta}\}_{\beta \in B}$. Since $\chi(\xi)$ is continuous in σ, for every $\varepsilon > 0$ there exist $\delta > 0$ and an inner product $(\ ,\ )_{\sigma\alpha}$ such that $\|\xi\|_{\sigma\alpha} < \delta$ implies $|\chi(0) - \chi(\xi)| < \varepsilon$. For this $\|\cdot\|_{\sigma\alpha}$, we can find $\|\cdot\|_{\tau\beta}$ such that $\|\cdot\|_{\sigma\alpha}$ is HS on E_0 with respect to $\|\cdot\|_{\tau\beta}$. Put

$$E'_\beta = \{f \in E^a \mid \text{restriction of } f \text{ on } E_0 \text{ is continuous in } \|\cdot\|_{\tau\beta}\},$$

then repeating the same discussion with the proof of Th.17.1, we see that $\mu^*(E'_\beta) \geq \chi(0)-\varepsilon$. ($\mu^*$ is the outer measure of μ). Since $E'_\beta \subset E'_\tau$, we have $\mu^*(E'_\tau) \geq \chi(0)-\varepsilon$, so letting $\varepsilon \to 0$ we get $\mu^*(E'_\tau)=\chi(0)$. Namely E'_τ is thick with respect to μ.

(q.e.d.)

Definition 18.4. Let (E,τ) be a Hilbertian type space. The Sazonov topology of τ is defined as the strongest Hilbertian type topology σ on E which is HS with respect to τ.

In other words, σ is defined by the family of all inner products which are HS with respect to some inner product in τ.

Theorem 18.3. Let (E,τ) be a metrizable Hilbertian type space, and suppose that μ lies on E'. Then the characteristic function of μ is continuous in the Sazonov topology of τ.

Corollary. Let (E,τ) be a metrizable Hilbertian type space. In order that μ lies on E', it is necessary and sufficient that the characteristic function of μ is continuous in the Sazonov topology of τ.

Proof (of Th.). Since (E,τ) is metrizable, it is defined by a countable family $\{\|\cdot\|_m\}$ of Hilbertian semi-norms. Put

$$E'_m = \{f \in E^a \mid f \text{ is continuous in } \|\cdot\|_m\},$$

then we have $E'=\bigcup_{m=1}^{\infty} E'_m$. Let U_m be the unit ball of E'_m, and B_{mn} be an equi-measure cover of nU_m. We have $\lim_{m\to\infty} \lim_{n\to\infty} \mu(B^c_{mn})=0$.

We shall define $(\xi,\eta)_{1m}$ by (18.1) replacing B_n by B_{mn}. Then, choosing n,m such that $\mu(B^c_{mn})<\varepsilon$, we see that $\|\xi\|_{1m} \leq \frac{\varepsilon}{n^2}$ implies $|\chi(0)-\chi(\xi)| \leq (2+\sqrt{\mu(E^a)})\varepsilon$. Evidently $\|\cdot\|_{1m}$

is HS with respect to $\|\cdot\|_m$. Therefore the Sazonov topology is stronger than the topology defined by $\{\|\cdot\|_{1m}\}$. Thus $\chi(\xi)$ is continuous in the Sazonov topology. (q.e.d.)

Remark. We need the assumption of metrizability for Th.18.3, but not for Th.18.2.

Example 18.1. We shall give a counter example of Th.18.3 for non-metrizable E. Let (H,τ) be a Hilbert space and σ be the Sazonov topology of τ. (H,σ) is a Hilbertian type space, and the dual space H'_σ is identical with $H'=H'_\tau$, because σ is stronger than the weak topology. The characteristic function of a measure μ on $H'=H'_\sigma$ is continuous in σ, but not always continuous in the Sazonov topology of σ. Let. $\|\cdot\|_1$ be a Hilbertian semi-norm which is HS with respect to τ, but not to σ. (For instance, for $H=(\ell^2)$, $\|\xi\|_1^2=\sum_{n=1}^{\infty}\frac{\xi_n^2}{n^2}$). Then $\chi(\xi)=\exp(-\|\xi\|_1^2/2)$ corresponds to a measure on $H'=H'_\sigma$, but it is not continuous in the Sazonov topology of σ.

§19. Supplementary results

Theorem 19.1. Let E be a vector space. A characteristic function $\chi(\xi)$ on E is constant on a subspace F, if and only if $F^\perp$ is a thick set in the corresponding measure μ, where

$$(19.1)\qquad F^\perp = \{f\in E^a \mid f(\xi) = 0 \quad \text{for} \quad {}^\forall\xi\in F\}.$$

Proof. For every $\xi\in F$, the set $\{f\in E^a \mid f(\xi)=0\}$ contains $F^\perp$, so that if $F^\perp$ is a thick set in μ, then the distribution of $f(\xi)$ concentrates in 0, therefore

$$\chi(\xi) = \int_{E^a} \exp(if(\xi))\, d\mu(f) = \mu(E^a) \quad (= \chi(0)).$$

Conversely, suppose that $\chi(\xi)$ is constant on F. It is sufficient to show that $B=\pi_M^{-1}(\pi_M(F^\perp))$ is thick for every countable dimensional $M\subset E$. $f\in B$ means that $\exists g\in F^\perp$, $\pi_M f=\pi_M g$, therefore it means that f is zero on $M\cap F$. Thus for a linear base $\{\xi_n\}$ of $M\cap F$, we have

$$(19.2)\quad B = \{f\in E^a \mid f(\xi_n) = 0 \text{ for } \forall n\} = \bigcap_{n=1}^{\infty}\{f\in E^a \mid f(\xi_n) = 0\}.$$

Since $\xi_n\in F$ implies $s\xi_n\in F$ for $\forall s\in\mathbb{R}$, we have $\chi(s\xi_n)=\chi(0)$, i.e.

$$\int_{E^a} \exp(isf(\xi_n))\, d\mu(f) = \chi(0) \quad \text{for } \forall s\in\mathbb{R}.$$

Therefore the distribution of $f(\xi_n)$ concentrates in 0, namely $\mu(\{f\in E^a \mid f(\xi_n)\neq 0\})=0$. Thus from (19.2) we get $\mu(B^c)=0$, hence B is a thick set with respect to μ. (q.e.d.)

Theorem 19.2. Let E be a vector space, E_0 be a subspace of E, and σ (resp. τ) be a Hilbertian type topology on E (resp. E_0). Suppose that $\chi(\xi)$ is constant on a subspace F of E, and that $\chi(\xi)$ is continuous in σ. If σ is HS on E_0 with respect to τ, then $E'_\tau\cap F^\perp$ is a thick set in the corresponding measure μ.

Proof. We shall prove that $\pi_M^{-1}(\pi_M(E'_\tau\cap F^\perp))$ is thick for every countable dimensional $M\subset E$. By Th.18.2, $\pi_M^{-1}(\pi_M(E'_\tau))$ is thick, and by Th.19.1, $\pi_M^{-1}(\pi_M(F^\perp))$ is a thick $\mathcal{B}_E$-measurable set. So that their intersection is a thick set. Thus the proof will be completed, if we show

$$(19.3)\qquad \pi_M^{-1}(\pi_M(E'_\tau\cap F^\perp)) = \pi_M^{-1}(\pi_M(E'_\tau))\cap \pi_M^{-1}(\pi_M(F^\perp)).$$

If f belongs to the right side of (19.3), then f is

continuous on $M\cap E_0$ in τ, and constantly zero on $M\cap F$. Let $\{\|\cdot\|_{\tau\beta}\}$ be the family of Hilbertian semi-norms defining the topology τ. We shall define a (non-Hausdorff) topology ρ on E_0+F by the family $\{\|\cdot\|_{\rho\beta}\}$, where

$$(19.4)\qquad \|\xi\|_{\rho\beta} = \inf\{\|\xi + \eta\|_{\tau\beta} \mid \eta\in F,\ \xi + \eta \in E_0\}.$$

Since $f(\xi)$ is continuous in ρ on $M\cap(E_0+F)$, we can extend it to a continuous linear function on E_0+F in ρ, then to a linear function g on E such that $g=f$ on M. Then, we have $\pi_M f=\pi_M g$ and $g\in E'_\tau\cap F$, because $\|\xi\|_{\rho\beta}\leq\|\xi\|_{\tau\beta}$ on E_0 and $\|\xi\|_{\rho\beta}=0$ on F. Thus f belongs to the left side of (19.3). (q.e.d.)

<u>Theorem 19.3</u>. Let E be a vector space, E_0 be a subspace of E, and τ be a metrizable Hilbertian type topology on E_0. If an inner product $(\ ,\ )$ is defined on E and is not HS on E_0 with respect to τ, then the function $\exp(-\|\xi\|^2/2)$ corresponds to a measure in which E'_τ is a null set, namely $\mu^*(E'_\tau)=0$.

<u>Corollary</u>. Let E be a vector space, E_0 be a subspace of E, and σ (resp. τ) be a Hilbertian type topology on E (resp. E_0). Assume that τ is metrizable. Then, if and only if σ is HS on E_0 with respect to τ, every continuous $\chi(\xi)$ in σ corresponds to a measure on E'_τ.

<u>Proof</u> (of Th.). Suppose that τ is defined by a countable family $\{\|\cdot\|_m\}$ of Hilbertian semi-norms, and put $E'_m=\{f\in E^a \mid$ restriction of f on E_0 is continuous in $\|\cdot\|_m\}$. Since $E'_\tau = \bigcup_{m=1}^{\infty} E'_m$, it is sufficient to show $\mu^*(E'_m)=0$. Since $\|\cdot\|$ is assumed not to be HS on E_0 with respect to $\|\cdot\|_m$, for

every $n>0$ there exists a finite ONS $\{\xi_{nk}\}_{k=1,2,\cdots,K(n)}$ in $(\ ,\ )_m$ such that $\sum_{k=1}^{K(n)} \|\xi_{nk}\|^2>n$. Take the union of such ONSs, and apply Schmidt's orthogonalization process to get a countable ONS $\{\xi_k\}$ in $(\ ,\ )_m$ such that

$$(19.5) \qquad \sum_{k=1}^{\infty} \|\xi_k\|^2 = \infty.$$

We shall calculate the integral:

$$(19.6) \qquad I_N = \int_{E^a} \exp(-\frac{1}{2}\sum_{k=1}^{N} f(\xi_k)^2)\ d\mu(f).$$

Taking the Fourier transform, we can rewrite I_N as follows:

$$I_N = (2\pi)^{-N/2}\int_{\mathbb{R}^N} \exp(-\frac{1}{2}\sum_{k=1}^{N} s_k^2)\chi(\sum_{k=1}^{N} s_k\xi_k)\ ds_1 ds_2\cdots ds_N$$

$$= (2\pi)^{-N/2}\int_{\mathbb{R}^N} \exp(-\frac{1}{2}\sum_{k=1}^{N} s_k^2 - \frac{1}{2}\sum_{j=1}^{N}\sum_{k=1}^{N} s_j s_k(\xi_j,\xi_k))\ ds_1 ds_2\cdots ds_N$$

$$= \prod_{k=1}^{N} (1+\lambda_k)^{-\frac{1}{2}},$$

where λ_k is the eigenvalue of the matrix $((\xi_j,\xi_k))$. Since $\prod_{k=1}^{N}(1+\lambda_k)\geqq 1+\sum_{k=1}^{N}\lambda_k$, we get

$$(19.7) \qquad I_N \leqq [1+\mathrm{Tr}((\xi_j,\xi_k))]^{-\frac{1}{2}} = (1+\sum_{k=1}^{N}\|\xi_k\|^2)^{-\frac{1}{2}}.$$

Letting $N\to\infty$, we see that $\lim_{N\to\infty} I_N=0$ because of (19.5). This means that

$$(19.8) \qquad \exp(-\frac{1}{2}\sum_{k=1}^{\infty} f(\xi_k)^2) = 0 \quad \text{for } \mu\text{-almost all } f.$$

On the other hand, $f\in E_m'$ implies $\sum_{k=1}^{\infty} f(\xi_k)^2<\infty$, hence $\exp(-\frac{1}{2}\sum_{k=1}^{\infty} f(\xi_k)^2)>0$. So comparing with (19.8), we see $\mu^*(E_m')=0$. (q.e.d.)

Remark. Th.19.3 is not valid if τ is non-metrizable. A counter example is provided by Example 18.1.

Example 19.1. Let $E=\mathbb{R}_0^\infty$, $E^a=\mathbb{R}^\infty$, and $\chi(\xi)=\exp(-\|\xi\|^2/2)$ where $\|\cdot\|$ is the usual (ℓ^2)-norm. In the corresponding gaussian measure μ, the set $(\ell^2)_a=\{x\in\mathbb{R}^\infty \mid \sum_{n=1}^\infty a_n^2 x_n^2<\infty\}$ is thick if $a\in(\ell^2)$, but a null set if $a\notin(\ell^2)$.

Though $(\ell^2)_a$ is thick and measurable for $a\in(\ell^2)$, we have

$$(19.9)\qquad \bigcap_{a\in(\ell^2)} (\ell^2)_a = (\ell^\infty),$$

which is a null set as shown in Example 17.1. We shall prove (19.9). If $x=(x_n)\notin(\ell^\infty)$, then $\forall k>0$, $\exists n(k)$, $|x_{n(k)}|>k$. Put $a_{n(k)}=1/k$, and choose a_n for $n\neq n(k)$, $k=1,2,\cdots$ so small to get $a=(a_n)\in(\ell^2)$. For this a, we have $\sum_{n=1}^\infty a_n^2x_n^2 \geq \sum_{k=1}^\infty a_{n(k)}^2 x_{n(k)}^2$ $\geq \sum_{k=1}^\infty \frac{1}{k^2}k^2=\infty$, hence $x\notin(\ell^2)_a$.

§20. Nuclear space

Definition 20.1. A topological vector space is said to be nuclear, if it is Hilbertian type and HS with respect to itself.

In other words, a nuclear space is a space whose topology is defined by a family $\{\|\cdot\|_\alpha\}$ of Hilbertian semi-norms such that $\forall\alpha\ \exists\alpha'$, $\|\cdot\|_\alpha$ is HS with respect to $\|\cdot\|_{\alpha'}$.

Applying Th.18.2, 18.3 and 19.3 to a nuclear space, we have:

Theorem 20.1. a) Let E be a nuclear space. If $\chi(\xi)$ is continuous on E, the corresponding measure lies on E', the topological dual space of E.

b) Let E be a metrizable nuclear space. Only if $\chi(\xi)$ is continuous on E, it corresponds to a measure on E'.

c) Let E be a metrizable Hilbertian type space. Only if E is nuclear, every continuous $\chi(\xi)$ on E corresponds to a measure on E'.

Counter example of b) and c) for non-metrizable E. Let H be a Hilbert space, σ be its Sazonov topology, and w be the weak topology. We have $H'=H'_\sigma=H'_w$. Though σ is not nuclear, every continuous $\chi(\xi)$ in σ corresponds to a measure on $H'=H'_\sigma$. Meanwhile, though w is nuclear, the characteristic function of a measure on $H'=H'_w$ is not always continuous in w.

In the following three sections, we shall study nuclear spaces, apart from the question of extension of measures. In this section, we shall give some examples of nuclear space.

Example 20.1. The space (s) of rapidly decreasing sequences.

$$(20.1) \qquad (s) = \{x = (x_n) \in \mathbb{R}^\infty \mid \lim_{n\to\infty} n^p x_n = 0 \quad \text{for } \forall p > 0\}.$$

The topology of (s) is defined by $\|\cdot\|_p$, where

$$(20.2) \qquad \|x\|_p = \operatorname{Max}_n |n^p x_n|.$$

But the topology of (s) is defined also by

$$(20.2)' \qquad \widetilde{\|x\|}_p^2 = \sum_{n=1}^\infty n^{2p} x_n^2,$$

because in general, for a real sequence $\{t_n\}$ we have $\operatorname{Max}_n |t_n| \leqq \sqrt{\sum_{n=1}^\infty t_n^2} \leqq \sqrt{\sum_{n=1}^\infty 1/n^2}\ \operatorname{Max}_n n|t_n|$.

Each $\widetilde{\|\cdot\|}_p$ is a Hilbertian norm, and $\widetilde{\|\cdot\|}_p$ is HS with respect to $\widetilde{\|\cdot\|}_{p+1}$, therefore (s) is a nuclear space.

This result can be generalized as follows: Let B be

a family of sequences of non-negative numbers, and put

(20.3) $\|x\|_b = \operatorname*{Sup}_n |b_n x_n|$, $\ell(B) = \{x \in \mathbb{R}^\infty \mid \|x\|_b < \infty$ for $\forall b \in B\}$.

The space $\ell(B)$ is Hausdorff, if and only if $\forall n \exists b \in B$, $b_n \neq 0$. It is nuclear under the condition $\forall b \in B \exists b' \in B$, $\sum_{n=1}^{\infty} (b_n/b'_n)^2 < \infty$ (but the term 0/0, if any, should be omitted from the summation).

Example 20.2. The space $(\mathscr{S})$ of rapidly decreasing C^∞-functions.

We shall explain the case of one real variable, but the results are valid also for the case of many variables.

For simplicity, we shall denote $\frac{d}{dt}$ with D. Put

(20.4) $(\mathscr{S}) = \{$real function φ on $\mathbb{R} \mid \varphi$ is infinitely differentiable and $\forall k, \ell$, $\|\varphi\|_{k\ell} \equiv \operatorname*{Max}_t |t^k D^\ell \varphi(t)| < \infty\}$.

The topology of $(\mathscr{S})$ is defined by $\{\|\varphi\|_{k\ell}\}$. By the relation $D(t\varphi) = \varphi + tD\varphi$, even if we change the order of t and D in the definition of $\|\varphi\|_{k\ell}$, we obtain the same topology on $(\mathscr{S})$.

Denote the L^2-norm and L^∞-norm with $\|\cdot\|_2$ and $\|\cdot\|_\infty$ respectively. Then we have $\|\varphi\|_{k\ell} = \|t^k D^\ell \varphi\|_\infty$. But the topology of $(\mathscr{S})$ is defined also by $\{\|t^k D^\ell \varphi\|_2\}$ as proved below.

First we have

$$\|\varphi\|_2^2 = \int_{-\infty}^{\infty} \varphi(t)^2 dt \leqq \int_{-\infty}^{\infty} \frac{dt}{(1+t^2)^2} \operatorname{Max}((1+t^2)\varphi(t))^2 = \frac{\pi}{2} \|(1+t^2)\varphi(t)\|_\infty^2.$$

Second, we have for $\varphi \in (\mathscr{S})$,

$$\varphi(t)^2 = 2\int_{-\infty}^{t} \varphi(t)\varphi'(t)dt \leqq 2\|\varphi\|_2 \|\varphi'\|_2 \leqq \|\varphi\|_2^2 + \|\varphi'\|_2^2,$$

hence we have $\|\varphi\|_\infty^2 \leqq \|\varphi\|_2^2 + \|D\varphi\|_2^2$. From these inequalities, $\{\|t^k D^\ell \varphi\|_\infty\}$ and $\{\|t^k D^\ell \varphi\|_2\}$ define the same topology on $(\mathscr{S})$. Thus $(\mathscr{S})$ is a Hilbertian type space.

Next, consider the linear operator $T: \varphi \to (t^2-D^2)\varphi$. The eigenvalue of T is $2n+1$ $(n=0,1,2,\cdots)$ and the corresponding eigen function is $H_n(t)\exp(-t^2/2)$ (where $H_n(t)$ is the Hermite polynomial). Since $\sum_{n=0}^{\infty} 1/(2n+1)^2<\infty$, T^{-1} is an HS operator so that $\|\varphi\|_2$ is HS with respect to $\|(t^2-D^2)\varphi\|_2$. In general, $\|t^kD^\ell\varphi\|_2$ is HS with respect to $\|(t^2-D^2)t^kD^\ell\varphi\|_2$. Thus $(\mathcal{S})$ is a nuclear space.

Example 20.3. The space $C^\infty(a,b)$ of all infinitely differentiable functions on the closed interval $[a,b]$. (Rigorously speaking, infinitely differentiable on (a,b) and every derivative converges as $t\to a+0$ and $t\to b-0$).

We shall define the semi-norm

(20.5) $$\|\varphi\|_k = \max_{a\leq t\leq b} |D^k\varphi(t)|,$$

and define the topology of $C^\infty(a,b)$ by $\{\|\varphi\|_k\}$. This topology is defined also by $\{\|D^k\varphi\|_2\}$ as proved below. First we have $\|\varphi\|_2^2=\int_a^b \varphi(t)^2dt\leq(b-a)\|\varphi\|_\infty^2$. Second, we use the relation $\varphi(t)=\varphi(t_0)+\int_{t_0}^t \varphi'(t)dt$ to get $\|\varphi\|_\infty\leq|\varphi(t_0)|+\sqrt{b-a}\,\|\varphi'\|_2$. Here we shall choose t_0 such that $\varphi(t_0)=\frac{1}{b-a}\int_a^b\varphi(t)dt$, then we have $|\varphi(t_0)|\leq\frac{1}{\sqrt{b-a}}\|\varphi\|_2$. Therefore we get $\|\varphi\|_\infty\leq\frac{1}{\sqrt{b-a}}\|\varphi\|_2+\sqrt{b-a}\,\|\varphi'\|_2$. Thus the topology of $C^\infty(a,b)$ is defined by $\{\|D^k\varphi\|_2\}$, so that $C^\infty(a,b)$ is a Hilbertian type space.

Since $C^\infty(a,b)$ is isomorphic to $C^\infty(0,\pi)$, we shall prove the nuclearity of $C^\infty(0,\pi)$. Let c_n be the expansion coefficient of $\varphi(t)$ with respect to a CONS $\{\frac{1}{\sqrt{\pi}}, \sqrt{\frac{2}{\pi}}\cos nt\}$, then we have $\|\varphi\|_2^2=\sum_{n=0}^{\infty} c_n^2$. For $n>0$, integrating by parts, we have

$c_n=\sqrt{\frac{2}{\pi}}\int_0^{\pi}\varphi(t)\cos nt\,dt=-\sqrt{\frac{2}{\pi}}\,\frac{1}{n}\int_0^{\pi}\varphi'(t)\sin nt\,dt$. Since $\{\sqrt{\frac{2}{\pi}}\sin nt\}$ is another CONS of $L^2(0,\pi)$, we get $\|\varphi'\|_2^2=\sum_{n=1}^{\infty}n^2c_n^2$. Thus in virtue of $\sum_{n=0}^{\infty}1/(n^2+1)<\infty$, we see that $\|\varphi\|_2$ is HS with respect to $\sqrt{\|\varphi\|_2^2+\|\varphi'\|_2^2}$. In general, $\|D^k\varphi\|_2$ is HS with respect to $[\|D^k\varphi\|_2^2+\|D^{k+1}\varphi\|_2^2]^{1/2}$. Therefore $C^{\infty}(0,\pi)$, hence $C^{\infty}(a,b)$ also, is a nuclear space.

§21. Heredity

Theorem 21.1. Every subspace, factor space (by a closed subspace) and the completion of a nuclear (resp. Hilbertian type) space is nuclear (resp. Hilbertian type).

Proof. If the topology of E is defined by a family $\{\|\cdot\|_{\alpha}\}$ of Hilbertian semi-norms, the topology of the subspace F is defined by the restriction of each $\|\cdot\|_{\alpha}$ on F. Evidently this restriction is Hilbertian on F. If $\|\cdot\|_{\alpha}$ is HS with respect to $\|\cdot\|_{\alpha'}$ on E, then so is on F.

Each $\|\cdot\|_{\alpha}$ is extended continuously on the completion $\bar{E}$. The topology of $\bar{E}$ is defined by the family of extended $\|\cdot\|_{\alpha}$, which is Hilbertian on $\bar{E}$. If $\|\cdot\|_{\alpha}$ is HS with respect to $\|\cdot\|_{\alpha'}$ on E, then so is on $\bar{E}$ as shown below: Let $\{\bar{\xi}_n\}_{n=1,2,\cdots,N}$ be a finite ONS in $\bar{E}$ with respect to $(\ ,\ )_{\alpha'}$, then we can find a sequence of ONSs $\{\xi_n^{(m)}\}$ in E such that $\lim_{m\to\infty}\xi_n^{(m)}=\bar{\xi}_n$. Then $\sum_{n=1}^{N}\|\xi_n^{(m)}\|_{\alpha}^2\leq M$ for $\forall m$ implies $\sum_{n=1}^{N}\|\bar{\xi}_n\|_{\alpha}^2\leq M$.

Let F be a subspace of E. The factor topology on E/F is defined as the strongest topology in which the canonical projection $E \to E/F$ is continuous. E/F is Hausdorff, if and only if F is closed in E. If the topology of E is defined by $\{\|\cdot\|_\alpha\}$, the factor topology is defined by $\{\|\cdot\|_{F\alpha}\}$, where

(21.1) $$\|\dot{\xi}\|_{F\alpha} = \inf\{\|\xi + \eta\|_\alpha \mid \eta \in F\}.$$

Suppose that $\|\cdot\|_\alpha$ is Hilbertian. Then $\|\dot{\xi}\|_{F\alpha}$ is equal to $\|P\xi\|_\alpha$, where P is the orthogonal projection to $F^\perp$. (Rigorously speaking, put $N=\{\xi\in E \mid \|\xi\|_\alpha=0\}$ and let H be the Hilbert space obtained by the completion of E/N, then P is the projection to $(F/(N\cap F))^\perp$ in H). Thus $\|\dot{\xi}\|_{F\alpha}$ is Hilbertian.

If $\|\cdot\|_{\alpha'}$ is HS with respect to $\|\cdot\|_\alpha$, then $\|\cdot\|_{F\alpha'}$ is HS with respect to $\|\cdot\|_{F\alpha}$ as shown below: If $\{\dot{\xi}_n\}$ is a finite ONS in $(\ ,\ )_{F\alpha}$, then $\{P\xi_n\}$ is an ONS in $(\ ,\ )_\alpha$, so that we have $\sum_n \|P\xi_n\|^2_{\alpha'} \leqq M$. Thus in virtue of $\|\dot{\xi}\|_{F\alpha'} \leqq \|P\xi\|_{\alpha'}$, we get $\sum_n \|\dot{\xi}_n\|^2_{F\alpha'} \leqq M$. ($P$ may not be an orthogonal projection with respect to $(\ ,\ )_{\alpha'}$, but $\|\dot{\xi}\|_{F\alpha'}$ is smaller than $\|P\xi\|_{\alpha'}$). (q.e.d.)

<u>Example 21.1</u>. Let $(\mathcal{D}_K)$ be the space of all C^∞-functions whose carriers are contained in K, where K is a given compact set. The semi-norm $\|\varphi\|_k$ is defined by $\|\varphi\|_k = \operatorname{Max}_{t\in K} \|D^k\varphi(t)\|$. Since the function t^k is bounded on K, $(\mathcal{D}_K)$ is a topological subspace of $(\mathcal{S})$, so that it is nuclear. (If $K\subset[a,b]$, $(\mathcal{D}_K)$ is also a topological subspace of $C^\infty(a,b)$).

<u>Theorem 21.2</u>. Let $\{E_\lambda\}_{\lambda\in\Lambda}$ be a family of nuclear (resp. Hilbertian type) spaces, then the direct product $\prod_\lambda E_\lambda$ is nuclear (resp. Hilbertian type).

<u>Proof</u>. Suppose that the topology of E_λ is defined by

$\{\|\cdot\|_{\lambda\alpha}\}_{\alpha\in A_\lambda}$, then the product topology is defined by $\{\widetilde{\|\cdot\|}_{\lambda\alpha}\}_{\lambda\in\Lambda,\alpha\in A_\lambda}$ where $\|\widetilde{(\xi_\lambda)}\|_{\lambda\alpha}=\|\xi_\lambda\|_{\lambda\alpha}$. If $\|\cdot\|_{\lambda\alpha}$ is Hilbertian, then $\widetilde{\|\cdot\|}_{\lambda\alpha}$ is Hilbertian. If $\|\cdot\|_{\lambda\alpha}$ is HS with respect to $\|\cdot\|_{\lambda\alpha'}$, then $\widetilde{\|\cdot\|}_{\lambda\alpha}$ is HS with respect to $\widetilde{\|\cdot\|}_{\lambda\alpha'}$.

(q.e.d.)

Remark. The family $\{\widetilde{\|\cdot\|}_{\lambda\alpha}\}_{\lambda\in\Lambda,\alpha\in A_\lambda}$ is not directed. To obtain a directed family defining the product topology, we must consider a semi-norm in the form $\|(\xi_\lambda)\|^2=\sum_n\|\xi_{\lambda_n}\|^2_{\lambda_n\alpha_n}$ which is a finite sum. But for this directed family also, the HS property is assured because:

Lemma 21.1. If $\|\cdot\|_{1n}$ is HS with respect to $\|\cdot\|_{2n}$ for $n=1,2,\cdots,N$, then putting $\|\xi\|_1^2=\sum_{n=1}^N\|\xi\|_{1n}^2$ and $\|\xi\|_2^2=\sum_{n=1}^N\|\xi\|_{2n}^2$, $\|\cdot\|_1$ is HS with respect to $\|\cdot\|_2$.

Proof. In general, we shall remark that if $\|\xi\|_1\leqq\|\xi\|_2\leqq\|\xi\|_3$ and $\|\cdot\|_1$ is HS with respect to $\|\cdot\|_2$, then $\|\cdot\|_1$ is HS with respect to $\|\cdot\|_3$, and that if $\|\cdot\|_{1n}$ is HS with respect to $\|\cdot\|_2$, then $\sqrt{\sum_{n=1}^N\|\cdot\|_{1n}^2}$ is HS with respect to $\|\cdot\|_2$.

Under the assumption of the lemma, $\|\xi\|_{2n}\leqq\|\xi\|_2$ implies that each $\|\cdot\|_{1n}$ is HS with respect to $\|\cdot\|_2$, so that $\|\cdot\|_1$ is HS with respect to $\|\cdot\|_2$.

This lemma can be generalized to:

Lemma 21.1'. If $\|\cdot\|_{1n}$ is HS with respect to $\|\cdot\|_{2n}$ for $n=1,2,\cdots$, then putting $\|\xi\|_1^2=\sum_{n=1}^\infty a_n^2\|\xi\|_{1n}^2$ and $\|\xi\|_2^2=\sum_{n=1}^\infty b_n^2\|\xi\|_{2n}^2$, $\|\cdot\|_1$ is HS with respect to $\|\cdot\|_2$ for sufficiently small a_n/b_n. Here we assume $\|\xi\|_1<\infty$ and $\|\xi\|_2<\infty$ for $^\forall\xi$.

Proof. Put

$$(2.1.2)\qquad M_n = \mathrm{Sup}\{\sum_j\|\xi_j\|_{1n}^2|\{\xi_j\} : \text{finite ONS in } (\ ,\ )_{2n}\}$$

then $|\xi|_2 \geq b_n \|\xi\|_{2n}$ implies that for every finite ONS $\{\xi_j\}$ in $(\ ,\ )_2$ we have $\sum_j \|\xi_j\|_{1n}^2 \leq M_n/b_n^2$. This holds for every n, therefore we get $\sum_j |\xi_j|_1^2 \leq \sum_{n=1}^{\infty} a_n^2 M_n/b_n^2$. So if $\sum_{n=1}^{\infty} a_n^2 M_n/b_n^2 < \infty$, then $\|\cdot\|_1$ is HS with respect to $\|\cdot\|_2$.

Definition 21.1. Let $\{E_\lambda\}_{\lambda\in\Lambda}$ be a family of locally convex spaces, and $\sum_\lambda E_\lambda$ be a subset of $\Pi_\lambda E_\lambda$ such that

$$(21.3)\qquad \sum_\lambda E_\lambda = \{(\xi_\lambda)\in \Pi_\lambda E_\lambda \,|\, \xi_\lambda = 0 \ \text{ except for finite number of } \lambda\}.$$

Putting $E_\lambda^0 = E_\lambda \times \prod_{\lambda'\neq\lambda} \{0\}$ for each λ, we see that $\sum_\lambda E_\lambda$ is spanned by $\{E_\lambda^0\}$. Each E_λ being isomorphic to E_λ^0, we can consider the strongest locally convex topology in which all the injections $E_\lambda \cong E_\lambda^0 \to \sum_\lambda E_\lambda$ are continuous. Imposing this topology, we call $\sum_\lambda E_\lambda$ the direct sum of $\{E_\lambda\}$.

If the topology of E_λ is defined by $\{\|\cdot\|_{\lambda\alpha}\}_{\alpha\in A_\lambda}$, then the direct sum topology is defined by $\{\|\cdot\|_{c(\cdot),\alpha(\cdot)}\}$, where

$$(21.4)\qquad \|(\xi_\lambda)\|_{c(\cdot),\alpha(\cdot)} = \sum_\lambda c(\lambda)\|\xi_\lambda\|_{\lambda,\alpha(\lambda)}.$$

Here $c(\cdot)$ runs over all positive functions on Λ, and $\alpha(\cdot)$ runs over all maps from Λ to A_λ for each $\lambda\in\Lambda$.

Theorem 21.3. The direct sum $\sum_{n=1}^{\infty} E_n$ of a countable family of nuclear (resp. Hilbertian type) spaces is nuclear (resp. Hilbertian type).

Proof. In virtue of the inequality: $\sum_{n=1}^{\infty} t_n^2 \leq (\sum_{n=1}^{\infty} |t_n|)^2 \leq \sum_{n=1}^{\infty} 1/n^2 \cdot \sum_{n=1}^{\infty} n^2 t_n^2$, the direct sum topology is defined also by $\{\widetilde{\|(\xi_n)\|}_{c(\cdot),\alpha(\cdot)}\}$, where

$$(21.5)\qquad \widetilde{\|(\xi_n)\|}^2_{c(\cdot),\alpha(\cdot)} = \sum_{n=1}^{\infty} c(n)^2 \|\xi_n\|^2_{n,\alpha(n)}.$$

If each $\|\cdot\|_{n,\alpha(n)}$ is Hilbertian, then so is $\widetilde{\|\cdot\|}_{c(\cdot),\alpha(\cdot)}$. If $\|\cdot\|_{n,\alpha(n)}$ is HS with respect to $\|\cdot\|_{n,\alpha'(n)}$ for each n, then by lemma 21.1', $\widetilde{\|\cdot\|}_{c(\cdot),\alpha(\cdot)}$ is HS with respect to $\widetilde{\|\cdot\|}_{c'(\cdot),\alpha'(\cdot)}$ for some $c'(\cdot)$. (q.e.d.)

Definition 21.2. Let $\{E_\lambda\}$ be a family of locally convex spaces parametrized by a directed set Λ. Suppose that for $\lambda'>\lambda$ a linear continuous surjection $p_{\lambda\lambda'}$ from $E_{\lambda'}$ to E_λ is defined and satisfies $p_{\lambda\lambda'}\circ p_{\lambda'\lambda''}=p_{\lambda\lambda''}$ for $\lambda<\lambda'<\lambda''$. Then the system $\{E_\lambda, p_{\lambda\lambda'}\}$ is called a projective family. The projective limit $\varprojlim E_\lambda$ is defined as follows: it is a topological subspace of $\prod_\lambda E_\lambda$ such that

(21.6) $$\varprojlim E_\lambda = \{(\xi_\lambda)\in \prod_\lambda E_\lambda \mid p_{\lambda\lambda'}(\xi_{\lambda'}) = \xi_\lambda \quad \text{for } \forall\lambda'>\lambda\}.$$

By Th.21.1 and Th.21.2, any projective limit of a family of nuclear (resp. Hilbertian type) spaces is nuclear (resp. Hilbertian type).

Example 21.2. The space $(\mathcal{E})$ of all C^∞-functions on the real line. The semi-norm is given by

(21.7) $$\|\varphi\|_{k\ell} = \max_{|t|\leq\ell} |D^k\varphi(t)|.$$

The space $(\mathcal{E})$ can be obtained as the projective limit of $C^\infty(-\ell,\ell)$, therefore $(\mathcal{E})$ is a nuclear space.

Definition 21.3. Let $\{E_\lambda\}$ be a family of locally convex spaces parametrized by a directed set Λ. Suppose that for $\lambda'>\lambda$ a linear continuous injection $p_{\lambda'\lambda}$ from E_λ into $E_{\lambda'}$ is defined and satisfies $p_{\lambda''\lambda'}\circ p_{\lambda'\lambda}=p_{\lambda''\lambda}$ for $\lambda<\lambda'<\lambda''$. Then the system $\{E_\lambda, p_{\lambda'\lambda}\}$ is called a inductive family. The inductive limit $\varinjlim E_\lambda$ is defined as follows: it is a topological factor space of $\sum_\lambda E_\lambda$ by a subspace $\bar{M}$, where

(21.8) $$M = \{(\xi_\lambda) \in \sum_\lambda E_\lambda \mid {}^\exists\lambda_0, \sum_\lambda p_{\lambda_0\lambda}(\xi_\lambda) = 0\}.$$

By Th.21.1 and Th.21.3, the inductive limit of a sequence of nuclear (resp. Hilbertian type) spaces is nuclear (resp. Hilbertian type).

Let $\{E_n\}$ be an inductive sequence. Identifying E_n with $p_{mn}(E_n) \subset E_m$, we get an increasing sequence $\{E_n\}$. Then as a set, $\varinjlim E_n$ can be identified with $\bigcup_{n=1}^\infty E_n$. The inductive limit topology is defined by

(21.9) $$\|\xi\|_{c(\cdot),\alpha(\cdot)} = \operatorname{Inf} \sum_{n=1}^\infty c(n) \|\xi_n\|_{n,\alpha(n)}, \quad \xi \in \bigcup_{n=1}^\infty E_n,$$

where $\xi = \sum_{n=1}^N \xi_n$ $(\xi_n \in E_n)$ and Inf is taken with respect to such representations of ξ. The sum in (21.9) can be replaced by $[\sum_{n=1}^\infty c(n)^2 \|\xi_n\|^2_{n,\alpha(n)}]^{1/2}$. (Rigorously speaking, $\varinjlim E_n$ is identified with $(\bigcup_{n=1}^\infty E_n)/N$, where N is the set of common zeros of the semi-norms $\|\cdot\|_{c(\cdot),\alpha(\cdot)}$).

<u>Example 21.3</u>. The space $(\mathcal{D})$ of all C^∞-functions with compact carriers. As a set, $(\mathcal{D}) = \bigcup_K (\mathcal{D}_K)$, or choosing $K=[-n,n]$, $(\mathcal{D}) = \bigcup_{n=1}^\infty (\mathcal{D}_{[-n,n]})$. The space $(\mathcal{D})$ is nuclear in the inductive limit topology of $(\mathcal{D}_{[-n,n]})$.

<u>Definition 21.4</u>. An inductive sequence $\{E_n, p_{mn}\}$ is said to be strict, if each p_{mn} is a homeomorphism from E_n into E_m.

In Example 21.3, $(\mathcal{D})$ is the strict inductive limit of $(\mathcal{D}_{[-n,n]})$.

<u>Theorem 21.4</u>. For a strict inductive sequence, the canonical injection $E_n \to \bigcup_{n=1}^\infty E_n \cong \varinjlim E_n$ is homeomorphic.

Proof. The injection $E_n \to \varinjlim E_n$ is continuous in general. It is homeomorphic, if for a given $\|\cdot\|_{n\alpha}(\alpha \in A_n)$, we can find $c(\cdot), \alpha(\cdot)$ such that $\|\xi\|_{n\alpha} \leqq \|\xi\|_{c(\cdot),\alpha(\cdot)}$ on E_n.

For $m<n$, since the injection $E_m \to E_n$ is continuous, there exist $c(m)$ and $\alpha(m)$ such that $\|\xi\|_{n\alpha} \leqq c(m)\|\xi\|_{m,\alpha(m)}$ on E_m. For $m=n$, put $c(n)=1$ and $\alpha(n)=\alpha$. For $m>n$, since $E_{m-1} \to E_m$ is homeomorphic, by mathematical induction we can choose $c(m)$ and $\alpha(m)$ such that $c(m-1)\|\xi\|_{m-1,\alpha(m-1)} \leqq c(m)\|\xi\|_{m,\alpha(m)}$ on E_{m-1}.

Now we shall prove that

$$(21.10) \quad \xi \in E_n,\ \xi = \sum_{m=1}^{N} \xi_m,\ \xi_m \in E_m \Rightarrow \|\xi\|_{n\alpha} \leqq \sum_{m=1}^{N} c(m)\|\xi_m\|_{m,\alpha(m)}.$$

If $N \leqq n$, from the choice of $c(m)$ and $\alpha(m)$ we have $\|\xi\|_{n\alpha} \leqq \sum_{m=1}^{N} \|\xi_m\|_{n\alpha} \leqq \sum_{m=1}^{N} c(m)\|\xi_m\|_{m,\alpha(m)}$. For $N>n$, $\xi = \sum_{m=1}^{N} \xi_m$ and $\xi \in E_n$ imply that ξ_N belongs to E_{N-1}, and we have $c(N-1)\|\xi_{N-1}\|_{N-1,\alpha(N-1)} + c(N)\|\xi_N\|_{N,\alpha(N)} \geqq c(N-1)\ (\|\xi_{N-1}\|_{N-1,\alpha(N-1)} + \|\xi_N\|_{N-1,\alpha(N-1)}) \geqq c(N-1)\|\xi_{N-1}+\xi_N\|_{N-1,\alpha(N-1)}$. Thus the inequality (21.10) can be obtained by mathematical induction. (q.e.d.)

§22. Dual space

Definition 22.1. Let E be a locally convex space defined by semi-norms $\{\|\cdot\|_\alpha\}$. A subset B of E is said to be bounded, if every $\|\cdot\|_\alpha$ is bounded on B.

This definition does not depend on the choice of the family of semi-norms. Every continuous linear function f is bounded on a bounded set.

In a normed space, the unit ball is bounded. Conversely

in a locally convex space E, if some neighbourhood of 0 is bounded, then the topology of E can be defined by a single norm, namely E is a normed space.

Definition 22.2. Let E be a locally convex space, and E' be its topological dual space. Putting $\|f\|'_B = \sup_{\xi \in B} |f(\xi)|$, the topology defined by $\{\|f\|'_B\}$ is called the strong (or dual) topology on E', where B runs over all bounded sets of E.

Theorem 22.1. If E is a metrizable nuclear (resp. Hilbertian type) space, then E' is nuclear (resp. Hilbertian type) in the strong topology.

Proof. Suppose that the topology of E is defined by a countable family $\{\|\cdot\|_n\}$ of Hilbertian semi-norms. Put

$$(22.1) \qquad \|\xi\|^2_{c(\cdot)} = \sum_{n=1}^{\infty} c(n)^2 \|\xi\|^2_n,$$

and $E_{c(\cdot)} = \{\xi \in E \mid \|\xi\|_{c(\cdot)} < \infty\}$.

$\|\cdot\|_{c(\cdot)}$ is a Hilbertian norm on $E_{c(\cdot)}$, and its unit ball $B_{c(\cdot)}$ is a bounded set of E. Conversely every bounded set B of E is contained in some $B_{c(\cdot)}$. Therefore, the strong topology of E' is defined by $\{\|f\|'_{c(\cdot)}\}$, where $\|\cdot\|'_{c(\cdot)} = \|\cdot\|'_{B_{c(\cdot)}}$ is the dual semi-norm of $\|\cdot\|_{c(\cdot)}$, thus E' is Hilbertian type.

Suppose that E is a nuclear space. Without loss of generality, we can assume that $\|\cdot\|_n$ is HS with respect to $\|\cdot\|_{n+1}$. By lemma 21.1', for any given $\|\cdot\|_{c(\cdot)}$ we can find $c'(\cdot)$ such that $\|\cdot\|_{c'(\cdot)}$ is HS with respect to $\|\cdot\|_{c(\cdot)}$ on $E_{c(\cdot)}$. Then the dual semi-norm $\|\cdot\|'_{c(\cdot)}$ is HS with respect to $\|\cdot\|'_{c'(\cdot)}$ by the following lemma.

Lemma 22.1. Let $\|\cdot\|$ and $\|\cdot\|_1$ be two Hilbertian semi-norms on E. If $\|\cdot\|_1$ is HS with respect to $\|\cdot\|$, then the dual norm $\|\cdot\|'$ is HS with respect to $\|\cdot\|_1'$ on E_1', where E_1' is the topological dual space of E in $\|\cdot\|_1$.

Proof. Put $N=\{\xi\in E \mid \|\xi\|=0\}$, and let H be the completion of E/N in $\|\cdot\|$. With some non-negative HS operator T, the semi-norm $\|\cdot\|_1$ is written as $\|\xi\|_1=\|T\xi\|$, as explained below Def. 17.2. Let $\{e_n\}\cup\{\varepsilon_\alpha\}$ be a CONS of H consisting of eigenvectors of T, where $Te_n=\lambda_n e_n$ $(\lambda_n>0)$ and $T\varepsilon_\alpha=0$. $f(\varepsilon_\alpha)$ is zero for $f\in E_1'$. Consider $f_k\in E_1'$ such that $f_k(e_n)=\lambda_k\delta_{kn}$, then $\{f_k\}$ is a CONS of E_1' and $\|f_k\|'=\lambda_k$, so that $\sum_{k=1}^{\infty}\lambda_k^2<\infty$ means that $\sum_{k=1}^{\infty}\|f_k\|'^2<\infty$ thus $\|\cdot\|'$ is HS with respect to $\|\cdot\|_1'$.

(q.e.d.)

Theorem 22.2. If E is a strict inductive limit of metrizable nuclear (resp. Hilbertian type) spaces, then E' is nuclear (resp. Hilbertian type) in the strong topology.

Proof. Let E be the strict inductive limit of metrizable Hilbertian type spaces $\{E_n\}$. By Th.21.4, E_n is homeomorphically imbedded in E, thus the closure $\bar{E}_n$ is also metrizable as a subspace of E.

Let B be a bounded set of E. Supposing the $B\subset\bar{E}_n$, we can find a bounded set B_1 of $\bar{E}_n$ such that $B\subset B_1$ and $\|\cdot\|_{B_1}'$ is Hilbertian. If E_n is nuclear, then $\bar{E}_n$ is nuclear, so that we can find another bounded set B_2 of $\bar{E}_n$ such that $\|\cdot\|_{B_2}'$ is Hilbertian and $\|\cdot\|_{B_1}'$ is HS with respect to $\|\cdot\|_{B_2}'$. Thus, we can reduce all discussions to the metrizable case using the following lemma.

Lemma 22.2. Let E be the strict inductive limit of $\{E_n\}$.

Then every bounded set of E is contained in some $\bar{E}_n$.

Proof. Suppose that B is not contained in any $\bar{E}_n$. Take an element $\xi_1 \in B$. Supposing $\xi_1 \in E_{n_1}$, consider a continuous linear function f_1 on E_{n_1} such that $f_1(\xi_1)=1$. Take another element $\xi_2 \in B \cap E_{n_1}^c$. Supposing $\xi_2 \in E_{n_2}$, extend f_1 to a continuous linear function f_2 on E_{n_2} satisfying $f_2(\xi_2)=2$. Repeat this procedure, and define step by step a continuous linear function f_k on E_{n_k} such that f_k is an extension of f_{k-1} and $f_k(\xi_k)=k$ for some $\xi_k \in B \cap E_{n_k}$. Then the sequence $\{f_k\}$ defines a linear function f on E. Since f is continuous on each E_{n_k}, it is continuous on E by the definition of the inductive limit topology. But f is not bounded on B, thus B is not a bounded set. (q.e.d.)

Example 22.1. $(\mathcal{D}')$; the space of all distributions in Schwartz's sense, $(\mathcal{S}')$; the space of slowly increasing distributions, and $(\mathcal{E}')$; the space of distributions of compact carriers.

They are the dual spaces of $(\mathcal{D})$, $(\mathcal{S})$ and $(\mathcal{E})$ respectively. $(\mathcal{S})$ and $(\mathcal{E})$ are metrizable nuclear spaces, while $(\mathcal{D})$ is a strict inductive limit of metrizable nuclear spaces. So that the spaces $(\mathcal{S}')$, $(\mathcal{E}')$ and $(\mathcal{D}')$ are nuclear spaces.

Appendix. Definition of nuclearity without using Hilbertian semi-norms

Besides our Def.20.1, there are some other definitions of a nuclear space. In these definitions, the nuclearity is defined for a locally convex space E, without assuming that it is Hilbertian type. But actually we can find a family of Hilbertian semi-norms defining the topology of E, and the definition becomes equivalent to ours.

The following definition comes from French school, especially due to Grothendieck.

Definition A.1. Let $\|\cdot\|_0$ and $\|\cdot\|_1$ be two semi-norms on E. $\|\cdot\|_0$ is said to be nuclear with respect to $\|\cdot\|_1$, if there exist sequence $\{x_n\}\subset E$ and $\{f_n\}\subset E^a$ such that

(A.1) $$\sum_{n=1}^{\infty} \|f_n\|_1' \|x_n\|_0 < \infty,$$

$$\forall x \in E, \quad x = \sum_{n=1}^{\infty} f_n(x)x_n \quad \text{in the semi-norm } \|\cdot\|_0.$$

Here $\|\cdot\|_1'$ is the dual norm of $\|\cdot\|_1$.

Definition A.2. Let E be a locally convex space defined by semi-norms $\{\|\cdot\|_\alpha\}$. E is called a nuclear space if

(A.2) $\forall\alpha\ \exists\alpha'$, $\|\cdot\|_\alpha$ is nuclear with respect to $\|\cdot\|_{\alpha'}$.

Note that this definition is independent of the choice of $\{\|\cdot\|_\alpha\}$ defining the topology of E.

Theorem A.1. Our Def.20.1 is equivalent to Def.A.2.

Proof. Assume that E is nuclear in the sense of Def.20.1. Let $\|\cdot\|_\alpha$ be a Hilbertian semi-norm, and $\|\cdot\|_{\alpha'}$, $\|\cdot\|_{\alpha''}$ be Hilbertian semi-norms such that $\|\cdot\|_\alpha$ is HS with respect to $\|\cdot\|_{\alpha'}$, while $\|\cdot\|_{\alpha'}$ is HS with respect to $\|\cdot\|_{\alpha''}$. We shall prove that $\|\cdot\|_\alpha$ is nuclear with respect to $\|\cdot\|_{\alpha''}$.

By the definition of HS property, $\sum_j \|x_j\|_\alpha^2$ is bounded when $\{x_j\}$ runs over all finite ONSs in $\|\cdot\|_{\alpha'}$. Denote this supremum with M. Applying similar discussions as (19.5), we can get a countable ONS $\{x_n\}$ in $\|\cdot\|_{\alpha'}$ such that $\sum_{n=1}^{\infty} \|x_n\|_\alpha^2 = M$. Then, we have $\forall x \in E$, $x = \sum_{n=1}^{\infty} (x,x_n)_{\alpha'} x_n$ in $\|\cdot\|_\alpha$, because putting $y_N = x - \sum_{n=1}^{N} (x, x_n)_{\alpha'} x_n$, we get $\sum_{n=1}^{N} \|x_n\|_{\alpha}^2 + \|y_N\|_{\alpha}^2 / \|y_N\|_{\alpha'}^2 \leq M$ so that $\|y_N\|_{\alpha}^2 / \|y_N\|_{\alpha'}^2 \to 0$ hence $\|y_N\|_{\alpha} \to 0$ as $N \to \infty$. Thus we have $x = \sum_{n=1}^{\infty} f_n(x) x_n$ where $f_n(x) = (x, x_n)_{\alpha'}$. Since $\sum_{n=1}^{\infty} \|f_n\|'_{\alpha''} \|x_n\|_{\alpha} \leq$ $(\sum_{n=1}^{\infty} \|f_n\|'^2_{\alpha''})^{1/2} (\sum_{n=1}^{\infty} \|x_n\|_{\alpha}^2)^{1/2}$, and since $\sum_{n=1}^{\infty} \|x_n\|_{\alpha}^2 = M$, the condition (A.1) is satisfied if $\sum_{n=1}^{\infty} \|f_n\|'^2_{\alpha''} < \infty$. However, by lemma 22.1 $\|\cdot\|'_{\alpha''}$ is HS with respect to $\|\cdot\|'_{\alpha'}$, and $\{f_n\}$ is an ONS in $(\ ,\ )'_{\alpha'}$.

Next we shall prove the converse. In (A.1), without loss of generality we can assume that $\|f_n\|'_1 = \|x_n\|_0$. First assuming that both $\|\cdot\|_0$ and $\|\cdot\|_1$ are Hilbertian, we shall show that $\|\cdot\|_0$ is HS with respect to $\|\cdot\|_1$.

Let $\{y_j\}$ be a finite ONS in $\|\cdot\|_1$, then we have

$$\|y_j\|_0 \leq \sum_{n=1}^{\infty} |f_n(y_j)| \|x_n\|_0 \leq (\sum_{n=1}^{\infty} f_n(y_j)^2)^{1/2} (\sum_{n=1}^{\infty} \|x_n\|_0^2)^{1/2},$$

hence $\sum_j \|y_j\|_0^2 \leq (\sum_{n=1}^{\infty} \|f_n\|'^2_1)(\sum_{n=1}^{\infty} \|x_n\|_0^2) < \infty$.

Thus if E is Hilbertian type, the condition (A.2) implies the nuclearity in the sense of Def.20.1. Therefore what we should prove is that all spaces satisfying (A.2) are Hilbertian type. This proof is provided by the following lemma.

Lemma A.1. Let $\|\cdot\|_0$ and $\|\cdot\|_1$ be two semi-norms on E. If the condition (A.1) is satisfied, we can find a Hilbertian semi-norm $\|\cdot\|_H$ such that

$$\text{(A.3)} \qquad \exists a > 0, \ \exists b > 0, \forall x \in E, a\|x\|_0 \leq \|x\|_H \leq b\|x\|_1.$$

Proof. Assume that $\|f_n\|'_1 = \|x_n\|_0$. We shall define $\|x\|_H$ by

$$\text{(A.4)} \qquad \|x\|_H^2 = \sum_{n=1}^{\infty} f_n(x)^2.$$

Then $\|\cdot\|_H$ is Hilbertian, and we have

$$\|x\|_0 \leqq \sum_{n=1}^{\infty} |f_n(x)|\,\|x_n\|_0 \leqq \|x\|_H \sqrt{\sum_{n=1}^{\infty} \|x_n\|_0^2},$$

and $\|x\|_H^2 \leqq \sum_{n=1}^{\infty} \|f_n\|_1'^2 \|x\|_1^2$ hence $\|x\|_H \leqq \|x\|_1 \sqrt{\sum_{n=1}^{\infty} \|f_n\|_1'^2}$

Thus putting $b=1/a=(\sum_{n=1}^{\infty} \|x_n\|_0^2)^{1/2}$, we get (A.3). (q.e.d.)

In a similar way, we have the following lemma.

<u>Lemma A.1'</u>. Instead of (A.1), if we assume $\sum_{n=1}^{\infty} c_n \|f_n\|_1' \|x_n\|_0 < \infty$ for some $\sum_{n=1}^{\infty} 1/c_n^2 < \infty$, then we can find two Hilbertian semi-norms $\|\cdot\|_{0H} \geqq a\|\cdot\|_0$ and $\|\cdot\|_{1H} \leqq b\|\cdot\|_1$ such that $\|\cdot\|_{0H}$ is HS with respect to $\|\cdot\|_{1H}$.

<u>Proof</u>. Again assume that $\|f_n\|_1' = \|x_n\|_0$. Put

(A.5) $$\|x\|_{0H}^2 = \sum_{n=1}^{\infty} \frac{1}{c_n} f_n(x)^2, \quad \|x\|_{1H}^2 = \sum_{n=1}^{\infty} c_n f_n(x)^2.$$

Evidently both $\|\cdot\|_{0H}$ and $\|\cdot\|_{1H}$ are Hilbertian, and $\|\cdot\|_{0H}$ is HS with respect to $\|\cdot\|_{1H}$. The inequalities $a\|x\|_0 \leqq \|x\|_{0H}$ and $\|x\|_{1H} \leqq b\|x\|_1$ can be proved in a similar way as in Lemma A.1.

(q.e.d.)

Grothendieck's definition of nuclearity has been studied mainly in connexion with the tensor product of locally convex spaces. For instance c.f. Grothendieck "Produits tensoriels topologiques et espaces nucléaires" Mem. Amer. Math. Soc. (1955).

Another style of the definition comes from Russian school, especially due to Mityagin. This definition has been considered in connexion with the approximation of a set by some finite dimensional set.

<u>Definition A.3</u>. Let E be a vector space, $\|\cdot\|$ be a semi-norm on E, U be the unit ball in $\|\cdot\|$, and A be a subset of

E. The following $d_n(A|U)$ is called the n-dimensional width of A with respect to $\|\cdot\|$.

(A.6) $$d_n(A|U) = \mathrm{Inf}\{c > 0 | A \subset E_{n-1} + cU$$

$$\text{for some } n\text{-}1\text{-dimensional subspace } E_{n-1} \text{ of } E\}.$$

If $A \subset A'$ and $U' \subset U$, we have $d_n(A|U) \leqq d_n(A'|U')$. If U' is a unit ball in some semi-norm, we have

(A.7) $$d_{2n}(A|U) \leqq d_n(A|U')d_n(U'|U).$$

<u>Definition A.4</u>. Let E be a locally convex space defined by semi-norms $\{\|\cdot\|_\alpha\}$. Denote the unit ball in $\|\cdot\|_\alpha$ with U_α. E is called a nuclear space, if for a positive number λ we have

(A.8) $$\forall\alpha\ \exists\alpha'\ \ d_n(U_{\alpha'}|U_\alpha) = O(\frac{1}{n^\lambda}).$$

Note that if (A.8) is satisfied for some $\lambda_0 > 0$, then (A.8) is satisfied for every $\lambda > 0$ in virtue of (A.7).

<u>Theorem A.2</u>. Our Def.20.1 is equivalent to Def.A.4.

<u>Proof</u>. Suppose that both $\|\cdot\|_\alpha$ and $\|\cdot\|_{\alpha'}$ are Hilbertian, and that $\|\cdot\|_\alpha$ is HS with respect to $\|\cdot\|_{\alpha'}$. We shall define a countable ONS $\{x_n\}$ in $\|\cdot\|_{\alpha'}$ by mathematical induction. Let E_n be the n-dimensional subspace spanned by $\{x_1, x_2, \cdots, x_n\}$. $(E_0 = \{0\})$. Putting $c_n = \sup\{\|x\|_\alpha | x \in E_n^\perp \cap U_{\alpha'}\}$, we shall choose $x_{n+1} \in E_n^\perp$ such that $\|x_{n+1}\|_{\alpha'} = 1$ and $\|x_{n+1}\|_\alpha \geqq c_n/2$. Since $\|\cdot\|_\alpha$ is HS with respect to $\|\cdot\|_{\alpha'}$, we have $\sum_{n=1}^\infty \|x_n\|_\alpha^2 < \infty$, so that $\sum_{n=1}^\infty c_n^2 < \infty$. It is evident that $d_{n+1}(U_{\alpha'}|U_\alpha) \leqq c_n$, hence we get $\sum_{n=1}^\infty d_n(U_{\alpha'}|U_\alpha)^2 < \infty$. Since $\{d_n(U_{\alpha'}|U_\alpha)\}$ is a

decreasing sequence, this implies that $d_n(U_{\alpha'}|U_\alpha)=O(\frac{1}{\sqrt{n}})$. Thus Def.20.1 implies Def.A.4 for $\lambda=1/2$.

Conversely, if $\lambda>3$, the condition $d_n(U_{\alpha'}|U_\alpha)=O(\frac{1}{n^\lambda})$ implies (A.1) (of course replacing 0 and 1 by α and α' respectively), so that Def.A.4 implies Def.A.2, hence Def.20.1. For the proof of this statement, we need the following lemma.

Lemma A.2. Every n-dimensional Banach space E has a biorthonormal system $\{x_j\}_{j=1,2,\cdots,n}\subset E$ and $\{f_k\}_{k=1,2,\cdots,n}\subset E^a$. This means that

$$f_k(x_j) = \delta_{kj}, \quad \|x_j\| = 1, \quad \|f_k\|' = 1. \tag{A.9}$$

Proof. Let $\{y_j\}_{j=1,2,\cdots,n}$ be a linear base of E, and consider the function

$$\Phi_{\{y_j\}}(g_1,g_2,\cdots,g_n) = |\det(g_k(y_j))| \tag{A.10}$$

on B'^n, where det means the determinant and B' is the unit ball in $\|\cdot\|'$. Since B' is compact, $\Phi_{\{y_j\}}$ attains the maximum at some $(f_1,f_2,\cdots,f_n)$. Even if we consider another base $\{z_j\}$ instead of $\{y_j\}$, the corresponding function $\Phi_{\{z_j\}}$ differs from (A.10) only by a constant factor, so that $\Phi_{\{z_j\}}$ attains its maximum also at $(f_1,f_2,\cdots,f_n)$.

Since the maximum value $\Phi_{\{y_j\}}(f_1,f_2,\cdots,f_n)>0$, $\{f_k\}$ is linearly independent, therefore there exist $x_1,x_2,\cdots,x_n\in E$ such that $f_k(x_j)=\delta_{kj}$. If we show that $\|x_j\|=1$ for $\forall j$, the proof will be completed.

First $f_1(x_1)=1$ implies $1\leq\|f_1\|'\|x_1\|=\|x_1\|$. On the other hand by Hahn-Banach's theorem, there exists $f\in B'$ such that $f(x_1)=\|x_1\|$. Then we have $\Phi_{\{x_j\}}(f,f_2,\cdots,f_n)=\|x_1\|$. Since

$\Phi_{\{x_j\}}(f_1,f_2,\cdots,f_n)=1$ is the maximum value on B'^n, this means $\|x_1\|\leq 1$. Combining these two inequalities, we get $\|x_1\|=1$. In a similar way, we get $\|x_2\|=\cdots=\|x_n\|=1$. (q.e.d.)

Remark. Suppose that E is infinite dimensional and that $\|\cdot\|$ is a semi-norm on E. For a finite dimensional subspace R of E, every $f\in R^a$ can be extended to a continuous linear function $\in E'$ of the same norm. Thus there exist $\{x_j\}\subset R$ and $\{f_k\}\subset E'$ such that $f_k(x_j)=\delta_{kj}$, $\|x_j\|=1$ and $\|f_k\|'=1$.

Now, we shall continue the proof of Th.A.2. Assume that $d_n(U_{\alpha'}|U_\alpha)=O(\frac{1}{n^\lambda})$. Then we have for some $c>0$

(A.11) $\forall n\ \exists E_{n-1}$(n-1-dimensional subspace of E), $U_{\alpha'}\subset E_{n-1}+\frac{c}{n^\lambda}U_\alpha$.

Let $\{x_k^{(n)}\}_{k=1,2,\cdots,n-1}$ and $\{f_k^{(n)}\}_{k=1,2,\cdots,n-1}$ be a biorthonormal system of E_{n-1} in $\|\cdot\|_\alpha$. Then putting $p_n(x)=\sum_{k=1}^{n-1}f_k^{(n)}(x)\cdot x_k^{(n)}$, we have $\|p_n\|_\alpha\leq\sum_{k=1}^{n-1}\|f_k^{(n)}\|'\|x_k^{(n)}\|_\alpha=n-1$.

We shall prove:

$$\text{(A.12)}\qquad \|x-p_n(x)\|_\alpha\leq\frac{c}{n^{\lambda-1}}\|x\|_{\alpha'}.$$

By (A.11), we have $\exists y\in E_{n-1}$, $\|x-y\|_\alpha\leq\frac{c}{n^\lambda}\|x\|_{\alpha'}$. Then we have $\|x-p_n(x)\|_\alpha\leq\|x-y\|_\alpha+\|y-p_n(x)\|_\alpha$, but since $y=p_n(y)$, $\leq\|x-y\|_\alpha+\|p_n(x-y)\|_\alpha\leq\|x-y\|_\alpha+(n-1)\|x-y\|_\alpha=n\|x-y\|_\alpha\leq\frac{c}{n^{\lambda-1}}\|x\|_{\alpha'}$. Thus we get (A.12).

If $\lambda>1$, (A.12) implies $x=\lim_{n\to\infty}p_n(x)$ which can be written in the sum form as follows:

$$\text{(A.13)}\qquad x=\sum_{n=1}^{\infty}(p_{n+1}(x)-p_n(x))\quad\text{in}\quad\|\cdot\|_\alpha.$$

Here $p_1(x)$ should be replaced by zero.

Let $\{y_k^{(n)}\}$ and $\{\varphi_k^{(n)}\}$ be a biorthonormal system of

$E_n \cup E_{n-1}$ in $\|\cdot\|_\alpha$, then we have

$$(A.14)\qquad x = \sum_{n=1}^{\infty}\sum_{k=1}^{2n-1} \varphi_k^{(n)}(p_{n+1}(x) - p_n(x))y_k^{(n)}.$$

Evidently $\|y_k^{(n)}\|_\alpha = 1$. On the other hand, $|\varphi_k^{(n)}(p_{n+1}(x)-p_n(x))| \leqq \|p_{n+1}(x)-p_n(x)\|_\alpha \leqq \|x-p_{n+1}(x)\|_\alpha + \|x-p_n(x)\|_\alpha \leqq \frac{2c}{n^{\lambda-1}}\|x\|_{\alpha'}$. Therefore putting $\psi_k^{(n)}(x) = \varphi_k^{(n)}(p_{n+1}(x)-p_n(x))$ we have $\|\psi_k^{(n)}\|'_{\alpha'} \leqq \frac{2c}{n^{\lambda-1}}$. So that if $\lambda>3$, we get

$$(A.15)\quad \sum_{n=1}^{\infty}\sum_{k=1}^{2n-1}\|\psi_k^{(n)}\|'_{\alpha'}\|y_k^{(n)}\|_\alpha \leqq \sum_{n=1}^{\infty}\frac{2(2n-1)c}{n^{\lambda-1}} \leqq 4c\sum_{n=1}^{\infty}\frac{1}{n^{\lambda-2}} < \infty.$$

Thus the condition (A.1) is satisfied for $\|\cdot\|_\alpha$ and $\|\cdot\|_{\alpha'}$.

(q.e.d.)

<u>Remark</u>. If the condition $d_n(U_{\alpha'}|U_\alpha) = O(\frac{1}{n^\lambda})$ is satisfied for $\lambda>4$, then we can find two Hilbertian semi-norms $\|\cdot\|_{0H}$ and $\|\cdot\|_{1H}$ such that $a\|x\|_\alpha \leqq \|x\|_{0H}$, $\|x\|_{1H} \leqq b\|x\|_{\alpha'}$, and $\|x\|_{0H}$ is HS with respect to $\|x\|_{1H}$. Because if $\lambda>4$, instead of (A.15) we have $\sum_{n=1}^{\infty}\sum_{k=1}^{2n-1} n^{1+\varepsilon}\|\psi_k^{(n)}\|'_{\alpha'}\|y_k^{(n)}\|_\alpha < \infty$ for some $\varepsilon>0$. So we can apply Lemma A.1', because $\sum_{n=1}^{\infty}\sum_{k=1}^{2n-1}\frac{1}{(n^{1+\varepsilon})^2} = \sum_{n=1}^{\infty}\frac{2n-1}{n^{2+2\varepsilon}} \leqq 2\sum_{n=1}^{\infty}\frac{1}{n^{1+2\varepsilon}} < \infty$.

NOTE

<u>Note for Chapter 2</u>. Extension theorem was originally formulated and proved for R^∞ by Kolmogorov.

A.N. Kolmogorov, "Grundbegriffe der Wahrscheinlichkeitsrechnung", Springer, 1933.

Later, the theorem was generalized to the case of the projective limit.

S. Bochner, "Harmonic analysis and the theory of probability", Univ. of California, 1955.

The theorems are most conveniently formulated for standard measurable spaces. For instance, c.f.

K.R. Parthasarathy, "Probability measures on metric spaces", Academic Press, 1967.

In this lecture note, the author has intended to formulate various extension theorems in a systematic way, choosing the classical Kolmogorov's theorem as the starting point. The notion of compact regular measures, which is a generalization of Radon measures, is introduced for this purpose.

For standard measurable spaces, the assumption of metrizability of Luzin spaces was removed in this lecture note. This is a progress comparing e.g. with Parthasarathy's book.

<u>Note for Chapter 3</u>. Sazonov and Minlos' theory appeared in late 50s.

V. Sazonov, "A remark on characteristic functionals" Theory Prob. Appl., vol.3 (1958) pp.188-192.

R.A. Minlos, "Generalized random processes and their

extension to measures", Trudy Moskov. Mat. Obšč. vol.8 (1959) pp.497-518.

Though Segal and Gross investigated similar problems independently of Russian school, they were contented with "weak distribution" and did not study the extension theorem explicitly.

I.E. Segal, "Tensor algebras over Hilbert space" I, Trans. Am. Math. Soc. vol.81 (1956) pp.106-134.

L. Gross, "Integration and non-linear transformations in Hilbert space" Trans. Am. Math. Soc. vol.94 (1960) pp.404-440.

In this lecture note, the author follows Sazonov and Minlos' formulations, but fairly improves them using the Hilbertian semi-norms more effectively.

The concept of nuclear space is introduced by French school in connection with the kernel theorem, the representation of a bilinear form by an integral kernel. For instance,

A. Grothendieck, "Produits tensoriels topologiques et espaces nucléaires" Mem. Am. Math. Soc. 1955.

Also Russian mathematicians, investigating approximation theories of function spaces and extension of measures, reached a similar notion. Eventually they noticed that it is equivalent to the nuclearity in the sense of Grothendieck.

B.S. Mityagin, "Approximate dimension and bases in nuclear spaces" Russian Math. Survey vol.16 No.4 (1961) pp.59-128.

A formulation using Hilbertian semi-norms appeared in

I.M. Gel'fand & N.Ja. Vilenkin, "Generalized functions" vol.4, Applications of harmonic analysis, (English trans.) Academic Press 1964.

But they assumed the metrizability (namely, σ-Hilbertian space). In this lecture note, removing the assumption of metrizability, the author presents more clearly the equivalency of different definitions of nuclearity in the Appendix.

Studies of nuclear spaces are also active in East Germany.

A. Pietsch, "Nuclear locally convex spaces" (English trans.) Springer 1972.

PART B

INVARIANCE AND QUASI-INVARIANCE OF MEASURES ON INFINITE DIMENSIONAL SPACES

PART B

Invariance and quasi-invariance of measures on infinite dimensional spaces

Introduction

Part B is rather independent of Part A, though the latter is quoted sometimes. This time our main concern is the study of invariance and quasi-invariance of measures on infinite dimensional spaces.

In Chapter 1, we shall study the invariant measure on a group, especially the famous Haar measure on a locally compact group and its inverse problem: Weil topology. An invariant or quasi-invariant measure exists on a measurable group only if the group admits a locally totally bounded topology. This result holds for vector spaces also. Thus, we conclude the non-existence of the invariant measure on any infinite dimensional vector space. Namely, let E be an infinite dimensional vector space, then no measure on E is invariant nor quasi-invariant under all translations $x \to x+y$ $(y \in E)$.

Therefore we must be contented with invariance or quasi-invariance under <u>some</u> translations. Let F be a subspace of E. A measure μ on E is said to be F-(quasi-)invariant if μ is (quasi-)invariant under $x \to x+y$ for $y \in F$. Since infinite measures are difficult to handle, and since no finite measure is invariant, we discuss mainly on quasi-invariance.

In Chapter 2, as a concrete and useful example of quasi-invariant measures, we shall study the gaussian measure. It is a measure on $\mathbb{R}^\infty$ and is (ℓ^2)-quasi-invariant. Some other properties of the gaussian measure will be investigated. (The gaussian measure can be defined also on a general infinite dimensional vector space).

Chapter 3 - Chapter 5 are devoted to the detailed dicussions on the invariance and quasi-invariance of measures on $\mathbb{R}^\infty$. Chapter 3 concerns rather a general discussion. Our question is: for a subspace F of $\mathbb{R}^\infty$, what is the condition for the existence of F-quasi-invariant or strictly F-quasi-invariant measure? Chapter 4 concerns the infinite product measures of one-dimensional measures. We shall discuss the condition for the quasi-invariance of measures in this type. Chapter 5 concerns the invariant measure on $\mathbb{R}^\infty$. We shall define infinite dimensional Lebesgue measure and investigate some of its properties. Our theory includes Moore-Hill theory as well as the linear transformations of a special one.

CHAPTER 1. INVARIANT MEASURE ON A GROUP

First we shall define a measurable group, and discuss a σ-finite invariant measure on it. The condition for its existence is stated in terms of the possibility of imposing a locally compact topology on the group. The invariant measure is unique if it exists.

§1. Measurable group, invariant and quasi-invariant measure

Definition 1.1. Let G be a group and $\mathcal{B}$ be a σ-algebra on it. $(G,\mathcal{B})$ is called measurable group, if $(x,y)\to xy^{-1}$ is a measurable map from $(G\times G,\mathcal{B}\times\mathcal{B})$ to $(G,\mathcal{B})$.

Here, $(G\times G,\mathcal{B}\times\mathcal{B})$ is the product measurable space of $(G,\mathcal{B})$.

It is easily seen that $x\to x^{-1}$, $x\to xy$, $x\to yx$ (for fixed y) are all measurable from $(G,\mathcal{B})$ to $(G,\mathcal{B})$.

Definition 1.2. Let $(G,\mathcal{B})$ be a measurable group, and μ be a measure on it. μ is said to be right-invariant if

$$(1.1) \qquad \forall E\in\mathcal{B},\quad \forall y\in G:\quad \mu(Ey)=\mu(E).$$

μ is said to be right-quasi-invariant if

$$(1.2) \qquad \mu(E)=0 \Longrightarrow \forall y\in G,\ \mu(Ey)=0.$$

We can define the left-invariance and left-quasi-invariance in a similar way.

Example 1.1. The following μ is right-invariant.

$$(1.3) \qquad \mu(E) \begin{cases} = \infty, \text{ if } E \text{ is an infinite set,} \\ = \text{number of elements of } E, \text{ if } E \text{ is a finite set.} \end{cases}$$

<u>Example 1.2.</u> Lebesgue measure on $\mathbb{R}^1$ is invariant.

Lebesgue measure on $\mathbb{R}^1$ is σ-finite, while the measure in Example 1.1 is not σ-finite on $\mathbb{R}^1$. They are not mutually absolutely continuous.

Hereafter, all measures are assumed to be σ-finite (without saying so explicitly). Under this assumption, the uniqueness of the invariant measure is ascertained (except a constant factor), but the existence is doubtful.

<u>Theorem 1.1.</u> On a measurable group $(G,\mathcal{B})$, every right-quasi-invariant measure is also left-quasi-invariant. Two right-quasi-invariant measures are mutually absolutely continuous. (Namely $\mu_1(E)=0$ implies $\mu_2(E)=0$, and vice versa).

<u>Proof</u>. Let μ be a right-quasi-invariant measure. First we shall prove that $\mu(E)=0$ implies $\mu(E^{-1})=0$. Put $F=\{(x,y);\ xy^{-1}\in E\}$, then by Fubini's theorem we have

$$(1.4)\qquad (\mu\times\mu)(F)=\int_G \mu(Ey)d\mu(y)=\int_G \mu(E^{-1}x)d\mu(x).$$

If $\mu(E)=0$, then $\mu(Ey)=0$ for all $y\in G$, therefore $(\mu\times\mu)(F)=0$, thus we get $\mu(E^{-1}x)=0$ for almost all x, hence $\mu(E^{-1})=0$.

Now the left-quasi-invariance of μ is proved as follows; $\mu(E)=0 \Rightarrow \mu(E^{-1})=0 \Rightarrow \mu(E^{-1}y)=0 \Rightarrow \mu(y^{-1}E)=0$.

Next, let μ_1 and μ_2 be two right-quasi-invariant measures. In (1.4) considering $(\mu_1\times\mu_2)(F)$ instead of $(\mu\times\mu)(F)$, we can prove that $\mu_1(E)=0 \Rightarrow \mu_2(E^{-1})=0$, hence $\mu_2(E)=0$. (q.e.d.)

<u>Theorem 1.2.</u> On a measurable group $(G,\mathcal{B})$, if a right invariant measure exists, then a left invariant measure also exists. Two right invariant measures are different only by a

constant factor. (Namely $\exists c>0$, $\forall E\in\mathcal{B}$, $\mu_2(E)=c\mu_1(E)$.)

Proof. Let μ be a right-invariant measure. Define a measure ν by the relation:

$$(1.5) \qquad \nu(E)=\mu(E^{-1}),$$

then ν is left-invariant, because $\nu(yE)=\mu(E^{-1}y^{-1})=\mu(E^{-1})=\nu(E)$.

Let μ_1 and μ_2 be two right-invariant measures. By Th. 1.1 they are mutually absolutely continuous, so by Radon-Nikodym's theorem we have

$$(1.6) \qquad \exists f(x)>0, \quad d\mu_2=fd\mu_1 \quad (\text{Namely } \mu_2(E)=\int_E f(x)d\mu_1(x)).$$

For the proof of the theorem, it is sufficient to show that $f(x)$ is constant almost everywhere.

From (1.6) we have $d\mu_{2,y}=f_y d\mu_{1,y}$, where $\mu_y(E)=\mu(Ey)$ and $f_y(x)=f(xy)$. Since $\mu_{2,y}=\mu_2$ and $\mu_{1,y}=\mu_1$, from the uniqueness of the density function we see $f(x)=f_y(x)$ for almost all x. Thus, putting $F=\{(x,y);\ f(x)\neq f(xy)\}$, we have $(\mu_1\times\mu_1)(F)=0$, therefore $\forall' x$, $\forall' y$, $f(x)=f(xy)$. (Here $\forall' x$ means "for almost all x"). Especially we have $\exists x_0$, $\forall' y$, $f(x_0)=f(x_0 y)$. Since μ_1 is left-quasi-invariant (by Th. 1.1), we have $\forall' y$, $f(x_0)=f(y)$, thus f is constant almost everywhere. (q.e.d.)

Theorem 1.3. On a measurable group $(G,\mathcal{B})$, if the right-invariant measure μ is finite, then μ is also left-invariant.

Proof. We shall prove $\mu(E)=\mu(E^{-1})$ for $\forall E\in\mathcal{B}$. Then μ is left-invariant as proved below (1.5).

Apply the right-invariance of μ to the equality (1.4), then we have $\mu(E)\mu(G)=\mu(E^{-1})\mu(G)$. Therefore if $0<\mu(G)<\infty$, we get $\mu(E)=\mu(E^{-1})$. (q.e.d.)

Example 1.3. Lebesgue measure on $\mathbb{R}^1$ is the unique invariant measure.

The multiplicative group $\mathbb{R}^1-\{0\}$ is also abelian, and its invariant measure is given by $d\mu = \frac{dx}{|x|}$.

Example 1.4. Let G be the set of all 2×2 regular matrices $(G=GL(2,\mathbb{R}))$. We shall denote its element with $\begin{pmatrix}\alpha & \beta\\ \gamma & \delta\end{pmatrix}$.

The measure:

$$(1.7) \qquad d\mu = \frac{d\alpha d\beta d\gamma d\delta}{\begin{vmatrix}\alpha & \beta\\ \gamma & \delta\end{vmatrix}^2}$$

is both right- and left-invariant, though μ is not finite.

Example 1.5. Let G be a subgroup of $GL(2,\mathbb{R})$ with the conditions $\gamma=0$ and $\delta=1$. $G = \left\{\begin{pmatrix}\alpha & \beta\\ 0 & 1\end{pmatrix};\ \alpha\neq 0\right\}$. Then $d\mu_1=\frac{d\alpha d\beta}{|\alpha|}$ is right-invariant but not left-invariant, while another measure $d\mu_2=\frac{d\alpha d\beta}{\alpha^2}$ is left-invariant but not right-invariant.

Theorem 1.4. On a measurable group $(G,\mathcal{B})$, if a right-quasi-invariant measure exists, then a right-invariant measure also exists.

Proof. First Step. Define a map T on $G\times G$ as follows:

$$(1.8) \qquad T\colon (x,y) \to (xy,y).$$

Then T is a measurable automorphism on $(G\times G, \mathcal{B}\times\mathcal{B})$.

We shall show that $\mu\times\mu$ is T-quasi-invariant.

For $F\in\mathcal{B}\times\mathcal{B}$, $(\mu\times\mu)(F)=0$ is equivalent to $\forall' y$, $\mu(F(y))=0$ where $F(y)=\{x;\ (x,y)\in F\}$. Similarly $(\mu\times\mu)(T(F))=0$ is equivalent to $\forall' y$, $\mu(T(F)(y))=0$. Since $T(F)(y)=F(y)y$, the right-quasi-invariance of μ implies the T-quasi-invariance of $\mu\times\mu$.

By Radon-Nikodym's theorem, we have

$$(1.9)\quad {}^{\prime}\ \exists f(x,y) > 0,\quad (\mu\times\mu)(T(F)) = \int_F f(x,y)d\mu(x)d\mu(y).$$

<u>Second Step</u>. Consider three maps on $G\times G\times G$ as follows:

$$T_1: (x,y,z) \to (xy,y,z),\quad T_2: (x,y,z) \to (x,yz,z),$$

$$T_3: (x,y,z) \to (xz,y,z).$$

Then, we can easily show $T_1\circ T_2=T_3\circ T_2\circ T_1$, therefore from the uniqueness of the density function we get

$$(1.10)\qquad \forall'(x,y,z),\quad f(x,yz)f(y,z)=f(xy,z)f(y,z)f(x,y).$$

Divide both hand sides by $f(y,z)$ to get

$$(1.11)\qquad \exists x_0,\quad \forall'(y,z),\quad f(x_0,yz)=f(x_0y,z)f(x_0,y).$$

Put $f(x_0,x_0^{-1}y)=g(y)$. Since μ is left-quasi-invariant, replacing x_0y by y we have

$$(1.12)\qquad \forall'(y,z),\quad f(y,z)=g(yz)/g(y).$$

<u>Third Step</u>. Putting $d\nu = \frac{1}{g}d\mu$, we shall prove that ν is right-invariant.

First, $\nu\times\mu$ is T-invariant, because from (1.9) and (1.12) we have $(\nu\times\mu)(T(F))=\int_{T(F)}\frac{1}{g(x)}d\mu(x)d\mu(y)= \int_F \frac{1}{g(xy)}f(x,y)d\mu(x)d\mu(y)$
$=\int_F \frac{1}{g(xy)}\frac{g(xy)}{g(x)}d\mu(x)d\mu(y) = \int_F \frac{1}{g(x)}d\mu(x)d\mu(y) = (\nu\times\mu)(F)$.

Apply this result to $F=A\times B$ ($A\in\mathcal{B}$, $B\in\mathcal{B}$), then we get

$$(1.13)\qquad \int_B \nu(Ay)d\mu(y) = \int_B \nu(A)d\mu(y) = \nu(A)\mu(B).$$

From the uniqueness of the density function (regarding the both hand sides of (1.13) as measures of B), we get

$$(1.14) \qquad \forall' y, \quad \nu(Ay)=\nu(A).$$

But actually (1.14) holds for $\forall y$ as proved below. Suppose that $\nu(A)=\alpha$, $\nu(Ay_0)=\beta$ and $\alpha\neq\beta$. Then putting $D=\{y\in G;\ \nu(Ay)=\alpha\}$ and $D_0=\{y\in G;\ \nu(Ay)=\beta\}$, we have $\mu(D^c)=\mu(D_0^c)=0$. Since $D\subset D_0^c$, we get $\mu(D^c)=\mu(D)=0$ which contradicts to $\mu(G)>0$. (q.e.d.)

§2. Haar measure on a locally compact group

Definition 2.1. Let G be a group, and τ be a Hausdorff topology on it. (G,τ) is called a topological group, if $(x,y)\to xy^{-1}$ is a continuous map from $(G\times G,\ \tau\times\tau)$ to (G,τ).

Here, $(G\times G,\ \tau\times\tau)$ is the product topological space of (G,τ).

It is easily seen that $x\to x^{-1}$, $x\to xy$, $x\to yx$ (for fixed y) are all continuous from (G,τ) to (G,τ).

On a topological group (G,τ), let $\mathcal{B}$ be the Borel field (=σ-algebra generated by all open sets), and $\mathbb{B}$ be the Baire field (=the smallest σ-algebra in which all continuous functions are measurable). Obviously $x\to x^{-1}$, $x\to xy$, $x\to yx$ (for fixed y) are all measurable with respect to both $\mathcal{B}$ and $\mathbb{B}$. But $(x,y)\to xy^{-1}$ is not necessarily measurable, so both $(G,\mathcal{B})$ and $(G,\mathbb{B})$ are not necessarily measurable groups.

If $\mathcal{B}\times\mathcal{B}=\mathcal{B}(\tau\times\tau)$ (=the Borel field of $\tau\times\tau$), then $(G,\mathcal{B})$ is a measurable group. Especially if (G,τ) is metrizable and separable, then $(G, \mathcal{B})$ is a measurable group according to Th. 11.2 c) in Part A.

If $(G, \mathcal{B})$ is not σ-separated, (especially if the cardinal number of G is more than the power of continuum), then the diagonal Δ is not $\mathcal{B}\times\mathcal{B}$-measurable according to Th. 7.1 in Part A, thus $(G,\mathcal{B})$ is not a measurable group. Note that even a compact group may have the cardinal number more than the power of continuum.

If $\mathbb{B}\times\mathbb{B} = \mathbb{B}(\tau\times\tau)$ (=the Baire field of $\tau\times\tau$), then $(G,\mathbb{B})$ is a measurable group. Especially if (G,τ) is locally compact and σ-compact, then $(G,\mathbb{B})$ is a measurable group according to Th. 11. 2 b) in Part A.

Definition 2.2. A Borel measure μ is said to be inner-regular if

$$\text{(2.1)} \qquad \forall O \text{ (open set)}, \quad \mu(O)=\sup_K\{\mu(K);\ K\subset O \text{ and } K \text{ is compact}\},$$

μ is said to be outer-regular if

$$\text{(2.2)} \qquad \forall E\in\mathcal{B}, \quad \mu(E) = \inf_O\{\mu(O);\ E\subset O \text{ and } O \text{ is open}\}.$$

Theorem 2.1. Let (G,τ) be a locally compact and σ-compact group. There exists a σ-finite, right-invariant measure μ on $(G,\mathcal{B})$. Such a measure μ is unique (except a constant factor) under the assumption of inner- and outer-regularity. The restriction of μ on $\mathbb{B}$ is also σ-finite.

Proof of uniqueness Suppose that μ is a σ-finite, right-invariant Borel measure which is inner- and outer-regular. Since μ is σ-finite and outer-regular, we have an open set

O such that $\mu(O)<\infty$. Since μ is right-invariant, we have $\mu(K)<\infty$ for every compact set K. Since G is locally compact and σ-compact, G can be written as a countable union of compact Baire sets, so that the restriction of μ on $\mathbb{B}$ is also σ-finite.

Since $(G, \mathbb{B})$ is a measurable group, Th. 1.2. assures that μ is uniquely determined on $\mathbb{B}$ except a constant factor. And the uniqueness on $\mathbb{B}$ implies the uniqueness on $\mathcal{B}$, as proved below. Since μ is inner-regular, we have (2.1). But G being locally compact, for every $K \subset O$ there exists a compact Baire set B which satisfies $K \subset B \subset O$. Therefore we have

$$(2.3) \qquad \mu(O) = \sup_B \{\mu(B);\ B \subset O \text{ and } B \text{ is a compact Baire set}\}.$$

Thus $\mu(O)$ is determined uniquely. Since μ is outer-regular, from (2.2) μ is determined uniquely on $\mathcal{B}$.

(q.e.d.)

For the proof of existence part of Th. 2.1, we need some preparations.

Definition 2.3. Let G be a topological group, and U be a symmetric neighbourhood of e (=identity element). A subset X of G is said to be U-separated if

$$(2.4) \qquad x \in X,\ x' \in X,\ x \neq x' \Rightarrow xx'^{-1} \notin U.$$

For a compact set K of G, put

$$(2.5) \qquad s(U,K) = \sup_X \{\#(X);\ X \subset K \text{ and } X \text{ is U-separated}\},$$

(2.6) $$r(U,K) = \inf_{Y}\{\#(Y);\ K\subset \bigcup_{y\in Y} Uy\},$$

where $\#(X)$ means the number of elements of X.

Lemma 2.1. Let U be a symmetric neighbourhood of e and V be another symmetric neighbourhood of e which satisfies $V^2\subset U$. Then we have $s(U,K)\leqq r(V,K)$.

(Since K is compact, we have $r(V,K)<\infty$, hence $s(U,K)<\infty$).

Proof. Suppose that $X\subset K$, X is U-separated and $K\subset \bigcup_{y\in Y} Vy$. Then we have $\#(X)\leqq \sum_{y\in Y} \#(X\cap Vy)$. However for every y we have $\#(X\cap Vy)\leqq 1$, because if $x,x'\in X\cap Vy$, then $xx'^{-1}\in Vyy^{-1}V^{-1} = V^2\subset U$, hence $x=x'$. Thus we get $\#(X)\leqq\#(Y)$, which implies $s(U,K)\leqq r(V,K)$. (q.e.d.)

Suppose that G is locally compact, and let U_0 be a fixed compact symmetric neighbourhood of e. We shall denote $s(U,U_0)$ with $s(U)$ and $r(U_0,K)$ with $r(K)$.

Now, for a compact set K we shall put

(2.7) $$\nu_U(K) = \frac{s(U,K)}{s(U)} .$$

Evidently we have

(2.8) $$\nu_U(K_1\cup K_2)\leqq\nu_U(K_1)+\nu_U(K_2).$$

If $K_1\cap K_2=\phi$, then we can find U such that $K_1K_2^{-1}\cap U=\phi$. For such a U, we have

(2.9) $$\nu_U(K_1\cup K_2) = \nu_U(K_1)+\nu_U(K_2).$$

Lemma 2.2. For every K we have $\nu_U(K)\leqq r(K)$.

Proof. Suppose that X is a U-separated set $\subset K$, and $K\subset \bigcup_{y\in Y} U_0y$. Then we have $\#(X)\leqq \sum_{y\in Y} \#(X\cap U_0y)$. However for every $y \in Y$ we have $\#(X \cap U_0 y) \leqq s(U)$, because $X \cap U_0 y = (Xy^{-1} \cap U_0)y$ and $Xy^{-1} \cap U_0$ is a U-separated set $\subset U_0$. Thus we get $\#(X) \leqq s(U)\#(Y)$, hence $s(U,K) \leqq s(U)r(K)$ which is equivalent to $\nu_U(K) \leqq r(K)$. (q.e.d.)

We want to define $\nu(K)$ by $\lim_{U \to \{e\}} \nu_U(K)$.

Let $\mathcal{C}$ be the family of all compact sets, and $\mathcal{U}$ be the family of all symmetric neighbourhoods of e. Define the map $\Phi: U \to (\nu_U(K))_{K \in \mathcal{C}}$ from $\mathcal{U}$ to $\prod_{K \in \mathcal{C}} [0,r(K)]$. By Tichonov's theorem $\prod_{K \in \mathcal{C}} [0,r(K)]$ is compact in the product topology, so that $\Phi(U)$ should have an accumulating point as $U \to \{e\}$. More precisely speaking, we have

$$(2.10) \qquad \bigcap_{U \in \mathcal{U}} \overline{\{\Phi(V);\ V \in \mathcal{U} \text{ and } V \subset U\}} \neq \phi.$$

Let $(\nu(K))_{K \in \mathcal{C}}$ be a point of this set. Then, we have

$$(2.11) \qquad \nu(K_1 \cup K_2) \leqq \nu(K_1) + \nu(K_2).$$

$$(2.12) \qquad \text{If } K_1 \cap K_2 = \phi, \text{ then } \nu(K_1 \cup K_2) = \nu(K_1) + \nu(K_2).$$

$$(2.13) \qquad \text{If } K_1 \subset K_2, \text{ then } \nu(K_1) \leqq \nu(K_2).$$

If X is U-separated, then evidently Xy is also U-separated. Thus $s(U,K) = s(U,Ky)$ for every $U \in \mathcal{U}$, hence we get

$$(2.14) \qquad \forall y \in G, \quad \nu(K) = \nu(Ky).$$

It is a routine work to define a Borel measure from this $\nu(K)$, which is called a content on $\mathcal{C}$.

<u>Theorem 2.2.</u> Suppose that G is a locally compact space and

that a content $\nu(K)$ is defined on $\mathcal{C}$ (=the family of all compact sets), satisfying (2.11) ∿ (2.13). Then, defining

$$(2.15) \quad \nu_*(O) = \sup_K\{\nu(K);\ K\subset O \text{ and } K \text{ is compact}\}$$

$$(2.16) \quad \nu^*(E) = \inf_O\{\nu_*(O);\ E\subset O \text{ and } O \text{ is open}\},$$

we get a Carathéodory's outer measure ν^*. Furthermore, every Borel set is measurable with respect to ν^*. Thus by Carathéodory's theorem, ν^* induces a σ-additive measure μ on the Borel field $\mathcal{B}$.

Before the proof of Th. 2.2, we shall complete the proof of existence in Th. 2.1. Since ν satisfies (2.14), the corresponding μ is evidently right-invariant. For an open set O, we have $\mu(O)=\nu_*(O)$, and for a compact set K, we have $\mu(K^0)\leqq\nu(K)\leqq\mu(K)$ where K^0 is the interior of K. Hence (2.15) implies (2.1) and (2.16) implies (2.2). Thus μ is inner-regular and outer-regular.

Assuming that G is σ-compact, we have $G=\bigcup_{n=1}^{\infty} K_n$ where K_n is a compact set. Since $\mu(K_n)<\infty$, μ is a σ-finite measure. (Since G is locally compact, a compact set K is contained in an open set O whose closure is compact, so that $\mu(K)\leqq\mu(O)\leqq\nu(\bar{O})<\infty$). Thus the proof of Th. 2.1 has been completed. (q.e.d.)

<u>Proof of Th. 2.2.</u> We shall check the axioms of Carathéodory's outer measure (cf. Definition 3.3 in Part A).

$$(2.17) \quad E_1\subset E_2 \Rightarrow \nu^*(E_1)\leqq\nu^*(E_2),$$

$$(2.18) \quad \nu^*\Big(\bigcup_{n=1}^{\infty} E_n\Big)\leqq\sum_{n=1}^{\infty}\nu^*(E_n).$$

(2.17) is evident from (2.16). If $E_n \subset O_n$ for each n, we have $\nu^*(\bigcup_{n=1}^{\infty} E_n) \leq \nu_*(\bigcup_{n=1}^{\infty} O_n)$, thus for the proof of (2.18) it is sufficient to show

$$(2.19) \qquad \nu_*(\bigcup_{n=1}^{\infty} O_n) \leq \sum_{n=1}^{\infty} \nu_*(O_n).$$

Suppose that $K \subset \bigcup_{n=1}^{\infty} O_n$. Each point x of K has a compact neighbourhood $U(x)$ which is contained in some O_n. Since K is compact, K is covered by a finite union of such $U(x)$. Therefore for some N we get $K \subset \bigcup_{n=1}^{N} K_n$, $K_n \subset O_n$ and K_n is compact. Hence we get $\nu(K) \leq \nu(\bigcup_{n=1}^{N} K_n) \leq \sum_{n=1}^{N} \nu(K_n) \leq \sum_{n=1}^{N} \nu_*(O_n) \leq \sum_{n=1}^{\infty} \nu_*(O_n)$. Thus (2.19) has been proved.

Finally we shall prove that an open set O is measurable with respect to ν^*, namely

$$(2.20) \qquad \forall E, \quad \nu^*(E) \geq \nu^*(E \cap O) + \nu^*(E \cap O^c).$$

From the definition (2.16), it is sufficient to show

$$(2.21) \qquad \forall O_1 \text{ (open set)}, \quad \nu_*(O_1) \geq \nu_*(O_1 \cap O) + \nu^*(O_1 \cap O^c).$$

Suppose that $K \subset O_1 \cap O$ and $K_1 \subset O_1 \cap K^c$. Then we have $\nu_*(O_1) \geq \nu(K \cup K_1) = \nu(K) + \nu(K_1)$. For a fixed K, take the supremum of the left hand side with respect to K_1, then we have $\nu_*(O_1) \geq \nu(K) + \nu_*(O_1 \cap K^c) \geq \nu(K) + \nu^*(O_1 \cap O^c)$. Again taking the supremum with respect to K, we have (2.21). (q.e.d.)

<u>Definition 2.4</u>. Let G be a locally compact and σ-compact group. The unique right-invariant measure on the Baire field $\mathbb{B}$ is called the right Haar measure on G. Sometimes its

unique extension on the Borel field $\mathcal{B}$ (under the condition of inner- and outer-regularity) is also called the right Haar measure on G.

The left Haar measure is defined in a similar way.

§3. Haar measure on a thick group

As for the uniform structure on a topological group G, we shall consider the right uniform structure.

Definition 3.1. In a topological group G, a set E is said to be precompact if

$$(3.1)\quad \forall U \text{ (neighbourhood of } e), \ \exists \{x_n\}_{n=1,2,\ldots,N}, \quad E \subset \bigcup_{n=1}^{N} Ux_n .$$

A topological group G is said to be locally precompact, if the identity element e has a precompact neighbourhood. G is said to be σ-precompact if G can be written as a countable union of precompact sets.

A compact set is evidently precompact. If G is complete, every precompact closed set is compact (for instance cf. N. Bourbaki: "Topologie générale").

Theorem 3.1. In order that G is locally precompact and σ-precompact, it is necessary and sufficient that G can be imbedded densely into a locally compact and σ-compact group $\hat{G}$.

Proof. For the proof of sufficiency, what we must prove is that for a compact set K of $\hat{G}$, $K \cap G$ is precompact in G.

Let U be a neighbourhood of e in G. Choose a neighbourhood $\hat{V}$ and $\hat{W}$ of e in $\hat{G}$ such that $\hat{V}\cap G\subset U$ and $\hat{W}\hat{W}^{-1}\subset\hat{V}$. Since K is compact, we have $K\subset\bigcup_{n=1}^{N}\hat{W}x_n$ for some $\{x_n\}\subset\hat{G}$. But G being dense in $\hat{G}$, we have $\exists y_n\in G$, $y_n\in\hat{W}x_n$, hence $\hat{W}x_n\subset\hat{W}\hat{W}^{-1}y_n\subset\hat{V}y_n$. This implies $K\cap G\subset\bigcup_{n=1}^{N}Uy_n$.

Conversely, suppose that G is locally precompact and σ-precompact. Let $\hat{G}$ be the completion of G. (Uniform structure of $\hat{G}$ is the completion of that of G, and the group operation in $\hat{G}$ is the continuous extension of that in G. For a topological group in general, the map $x\to x^{-1}$ can not be extended continuously on $\hat{G}$, but under the assumption of locally precompactness, the map $x\to x^{-1}$ has a unique continuous extension on $\hat{G}$. cf. N. Bourbaki: "Topologie génèrale"). Then we can easily see that if E is precompact in G, the closure of E in $\hat{G}$ is compact. From this, we see that $\hat{G}$ is locally compact and σ-compact. (q.e.d.)

Theorem 3.2. Let G be a locally precompact and σ-precompact group, and define $\mathbb{B}_u$ as the smallest σ-algebra in which all uniformly continuous functions on G are measurable. Let $\hat{G}$ be the completion of G and $\hat{\mathbb{B}}$ be the Baire field of $\hat{G}$. Then we have $\mathbb{B}_u=\hat{\mathbb{B}}\cap G$, and $(G,\mathbb{B}_u)$ is a measurable group.

Proof. Let f be uniformly continuous on G, then its continuous extension $\hat{f}$ on $\hat{G}$ is $\hat{\mathbb{B}}$-measurable, therefore f is $\hat{\mathbb{B}}\cap G$-measurable, hence we get $\mathbb{B}_u\subset\hat{\mathbb{B}}\cap G$.

Conversely let g be continuous on $\hat{G}$. Since $\hat{G}$ is locally compact and σ-compact, g can be expressed as $g(x)=\lim_{n\to\infty}g_n(X)$, each g_n being a uniformly continuous function. So that the restriction of g on G is $\mathbb{B}_u$-measurable, hence we get $\hat{\mathbb{B}} \cap G \subset \mathbb{B}_u$.

Let $\hat{\Phi}$ be the map $(x,y) \to xy^{-1}$ defined on $\hat{G} \times \hat{G}$, and Φ be its restriction on $G \times G$. For $E \in \hat{\mathbb{B}}$ we have $\Phi^{-1}(E \cap G) = \hat{\Phi}^{-1}(E) \cap (G \times G)$. Since $(\hat{G}, \hat{\mathbb{B}})$ is a measurable group we have $\hat{\Phi}^{-1}(E) \in \hat{\mathbb{B}} \times \hat{\mathbb{B}}$, hence $\Phi^{-1}(E \cap G) \in (\hat{\mathbb{B}} \times \hat{\mathbb{B}}) \cap (G \times G)$. So the relation $(\hat{\mathbb{B}} \times \hat{\mathbb{B}}) \cap (G \times G) = (\hat{\mathbb{B}} \cap G) \times (\hat{\mathbb{B}} \cap G)$ assures that $(G, \mathbb{B}_u)$ is a measurable group. (q.e.d.)

<u>Definition 3.2</u>. A topological group G is said to be thick, if it satisfies:

(1) G is locally precompact and σ-precompact, and

(2) G is thick with respect to the Haar measure $\hat{\mu}$ on $\hat{G}$, namely

$$(3.2) \qquad B \in \hat{\mathbb{B}}, \quad B \cap G = \phi \Rightarrow \hat{\mu}(B) = 0.$$

<u>Theorem 3.3</u>. Let G be a thick group. There exists a right-invariant measure on the measurable group $(G, \mathbb{B}_u)$. (The right-invariant measure is unique except a constant factor according to Th. 1.2. This unique measure is called the right Haar measure on G).

<u>Proof</u>. Let μ be the trace of $\hat{\mu}$ on G, namely

$$(3.3) \qquad \mu(E) = \hat{\mu}(B), \quad E = B \cap G, \quad B \in \hat{\mathbb{B}}.$$

Because of thickness of G, we can define μ consistently by (3.3). μ is right-invariant, because for $y \in G$ we have $\mu(Ey) = \mu((B \cap G)y) = \mu(By \cap G) = \hat{\mu}(By) = \hat{\mu}(B) = \mu(E)$. The σ-finiteness of μ implies that of μ, because $\hat{G} = \bigcup_{n=1}^{\infty} B_n$, $\hat{\mu}(B_n) < \infty$ implies

$$G=\bigcup_{n=1}^{\infty}(B_n\cap G),\ \mu(B_n\cap G)<\infty. \qquad \text{(q.e.d.)}$$

Example 3.1. Example of a thick group.

Let T be a compact group, and G be non-countable direct product of T. G is a compact group in the product topology. Consider the following subgroup G_0:

(3.4) $\quad G_0=\{x=(x_\lambda)_{\lambda\in\Lambda}\in G;\ x_\lambda=e$ except countable number of $\lambda\in\Lambda\}$.

The Baire field of G is the direct product of that of T (cf. Th. 11.2b) in Part A), so that every $B\in\mathbb{B}$ depends only on countable number of coordinates, thus $B\cap G_0=\phi$ implies $B=\phi$. This shows that G_0 is a thick group.

The Haar measure on G is the direct product of that on T, and its trace on G_0 is the Haar measure on G_0.

Example 3.2. Let $\mathbb{Q}$ be the set of all rational numbers. Consider two topologies on $\mathbb{Q}$: τ is the topology induced from the real line $\mathbb{R}$, and τ' is the discrete topology. Both topologies τ and τ' make $\mathbb{Q}$ locally precompact and σ-precompact. $\mathbb{Q}$ is thick in τ', but not in τ.

Consider the smallest σ-algebra $\mathbb{B}_u(\tau)$ (resp. $\mathbb{B}_u(\tau')$) in which all uniformly continuous functions are measurable. Then we have $\mathbb{B}_u(\tau)=\mathbb{B}_u(\tau')$=the σ-algebra of all subsets of $\mathbb{Q}$. The Haar measure on $\mathbb{Q}$ in the topology τ' is given by $\mu(E)$ =number of elements of E.

Note that (1) the existence of invariant measure on $(G,\mathbb{B}_u(\tau))$ does not imply the thickness of G in the topology τ, and (2) $\mathbb{B}_u(\tau)=\mathbb{B}_u(\tau')$ does not imply $\tau=\tau'$.

But the following Th. 3.4 and Th. 3.5 relieve us from such unpleasant situations.

Definition 3.3. Let G be a locally precompact and σ-precompact group. A measure μ on $\mathbb{B}_u$ is said to be locally finite, if it satisfies

$$(3.5) \qquad \forall B \in \mathbb{B}_u, \ B \text{ is precompact} \Longrightarrow \mu(B) < \infty.$$

The Haar measure on a thick group is evidently locally finite.

Theorem 3.4. Let G be a locally precompact and σ-precompact group. If a right-invariant, locally finite measure μ exists on $(G, \mathbb{B}_u)$, then G is a thick group and μ is the right Haar measure on G.

Proof. We have only to prove that G is a thick group. (If G is a thick group, μ must be the Haar measure because of the uniqueness of right-invariant measure).

Let $\hat{G}$ be the completion of G, $\hat{\mathbb{B}}$ be the Baire field of $\hat{G}$, and define a measure $\hat{\mu}$ on $\hat{\mathbb{B}}$ as follows:

$$(3.6) \qquad \hat{\mu}(B) = \mu(B \cap G).$$

Then G is evidently thick with respect to $\hat{\mu}$. So, what we must prove is that $\hat{\mu}$ is the Haar measure on $\hat{G}$.

$\hat{\mu}$ is obviously σ-finite, because μ is so. $\hat{\mu}$ is right-invariant with respect to $y \in G$, because $\hat{\mu}(By) = \mu(By \cap G) = \mu((B \cap G)y) = \mu(B \cap G) = \hat{\mu}(B)$. Since G is dense in $\hat{G}$, the following lemma completes the proof.

Lemma 3.1. Let G be a locally compact and σ-compact group, $\mathbb{B}$ be its Baire field, and μ be a locally finite measure on $(G, \mathbb{B})$. Then, the set:

(3.7) $\quad G_\mu=\{y\in G;\ ^\forall B\in\mathbb{B},\ \mu(B)=\mu(By)\}$

is a closed subgroup of G.

Proof. G_μ is evidently a group. The proof of the closedness of G_μ is as follows.

First Step. If μ is a finite measure on $(G,\mathbb{B})$, it is regular in the sense that for $^\forall B\in\mathbb{B}$ we have

(3.8) $\quad ^\forall\varepsilon>0,\ ^\exists O$ (open Baire), $^\exists K$ (compact Baire),

$$K\subset B\subset O \text{ and } \mu(O)-\mu(K)<\varepsilon.$$

Let $\mathfrak{A}$ be the family of all sets $B\in\mathbb{B}$ which satisfy (3.8). Using the fact that G is locally compact and σ-compact, we can easily show that $\mathfrak{A}$ becomes a σ-algebra. If f is a continuous function on G, we have $f^{-1}((-\infty,a))=\bigcup_{n=1}^{\infty}f^{-1}((-\infty,a-\frac{1}{n}])$ for any $a\in\mathbb{R}$. This implies $f^{-1}((-\infty,a))\in\mathfrak{A}$, hence we get $\mathfrak{A}=\mathbb{B}$.

Second Step. Even if μ is locally finite, (3.8) holds for a precompact Baire set B. Let $O\supset B$ be a precompact open Baire set, and μ_0 be the restriction of μ on O: $\mu_0(E)=\mu(O\cap E)$ for $^\forall E\in\mathbb{B}$. Since μ_0 is a finite measure, we can apply First Step to μ_0. From this we get (3.8) for μ.

Third Step. For a precompact open Baire set O, $\mu(Oy)$ is a lower semi-continuous function of y. It is sufficient to prove this only at $y=e$. For a given $\varepsilon>0$, choose K such that $K\subset O$ and $\mu(O)-\mu(K)<\varepsilon$. Then there exists a symmetric neighbourhood U of e such that $KU\subset O$, in other words for $y\in U$ we have $K\subset Oy$. Hence we have $\mu(Oy)\geq\mu(K)>\mu(O)-\varepsilon$. Thus $\mu(Oy)$ is lower semi-continuous at $y=e$.

<u>Fourth Step</u>. Put G'_μ as follows:

(3.9) $G'_\mu=\{y\in G;\ ^\forall O$ (precompact open Baire set) $\mu(Oy)\leqq\mu(O)\}$.

Since $\mu(Oy)$ is lower semi-continuous, G'_μ is a closed set. Furthermore $y\in G'_\mu\cap G'^{-1}_\mu$ implies $\mu(Oy)\leqq\mu(O)$ and $\mu(Oyy^{-1})\leqq\mu(Oy)$, hence $\mu(O)=\mu(Oy)$.

We shall prove $G'_\mu\cap G'^{-1}_\mu=G_\mu$. Supposing $y\in G'_\mu\cap G'^{-1}_\mu$, we shall prove that $^\forall B\in\mathbb{B}$, $\mu(B)=\mu(By)$. From the Second Step, if B is precompact, we know that $\mu(B)$ is the infimum of $\mu(O)$, O being a precompact open Baire set containing B. From this we get $\mu(B)=\mu(By)$. Even if B is not precompact, since G is locally compact and σ-compact, B can be written as a countable union of precompact Baire sets. Thus we get $\mu(B)=\mu(By)$. (q.e.d.)

<u>Theorem 3.5</u>. Suppose that G is a thick group both in topologies τ and τ'. Then $\mathbb{B}_u(\tau)=\mathbb{B}_u(\tau')$ implies $\tau=\tau'$.

<u>Proof</u> will be completed, if a fundamental system of neighbourhoods of e (in τ or τ') can be expressed in terms of the right Haar measure. The following theorem will suffice this purpose.

<u>Theorem 3.6</u>. Let G be a thick group in the topology τ, and μ be the right Haar measure on $(G,\mathbb{B}_u)$. Put

(3.10) $\mathfrak{A}=\{E\in\mathbb{B}_u;\ 0<\mu(E)<\infty\}$,

(3.11) $W_{E,\varepsilon}=\{x\in G;\ \mu(E\Delta Ex)<\varepsilon\}$, then $\{W_{E,\varepsilon}\}_{E\in\mathfrak{A},\varepsilon>0}$ becomes a fundamental system of neighbourhoods of e in the given topology τ.

Proof. Let $\mathfrak{U}$ be the family of all symmetric, precompact neighbourhoods of e. Since $\mathfrak{U}\cap\mathbb{B}_u$ becomes a fundamental system of neighbourhoods of e, for any given $U\in\mathfrak{U}$, we can find $V\in\mathfrak{U}\cap\mathbb{B}_u$ such that $V^2\subset U$. Then, $\mu(V)<\infty$ because V is precompact, and $\mu(V)>0$ because V is a neighbourhood of e, thus we see $V\in\mathfrak{A}$. If $\varepsilon<2\mu(V)$, then $x\in W_{V,\varepsilon}$ implies $V\cap Vx\neq\phi$, hence $x\in V^{-1}V=V^2$. Thus we get $W_{V,\varepsilon}\subset V^2\subset U$.

Conversely, we shall show that any $W_{E,\varepsilon}$ in the form of (3.11) contains a neighbourhood of e. Let $\hat{G}$ be the completion of G, $\hat{\mathbb{B}}$ be the Baire field of $\hat{G}$, and $\hat{\mu}$ be the right Haar measure on $\hat{G}$. Every $E\in\mathfrak{A}$ can be written as $E=B\cap G$, $B\in\hat{\mathbb{B}}$, $0<\hat{\mu}(B)<\infty$. For $x\in G$, we have $E\Delta Ex=(B\Delta Bx)\cap G$ so that $\mu(E\Delta Ex)=\hat{\mu}(B\Delta Bx)$. So it is sufficient to prove the theorem for $\hat{G}$, or in other words we can assume that G is locally compact and σ-compact.

Under this assumption, $B\in\mathfrak{A}$ implies that (3.8) holds for B. (In Second Step of Lemma 3.1, the assumption of precompactness of B can be replaced by the condition $\mu(B)<\infty$. The proof requires a little modification). For K and O in (3.8), there exists a symmetric neighbourhood U of e such that $KU\subset O$. But $Ky\subset O$ implies $\mu(B\cap(By)^c)\leq\mu(O\cap(Ky)^c)=\mu(O)-\mu(Ky)=\mu(O)-\mu(K)<\varepsilon$. In a similar way, $K\subset Oy$ implies $\mu(By\cap B^c)<\varepsilon$. Thus $y\in U$ implies $\mu(B\Delta By)<2\varepsilon$, namely we have $U\subset W_{B,2\varepsilon}$. (q.e.d.)

Example 3.3. Let G and G_0 be of the same meaning as in Example 3.1. Take an element $x\in G$ such that $x^n\notin G_0$ for any $n \neq 0$, then the group G_1 generated by G_0 and x is isomorphic to $G_0 \times \mathbb{Z}$. Consider the following two topologies τ and τ' on G_1: τ is the topology of a topological subgroup of G, while τ' is the product topology of G_0 and $\mathbb{Z}$, G_0 being a topological subgroup of G and $\mathbb{Z}$ being a discrete group. τ' is stronger than τ. G_0 is dense in G_1 in the topology τ, but not in the topology τ'. Thus we see $\tau \neq \tau'$.

The completion of G_1 is G in τ, and $G \times \mathbb{Z}$ in τ'. Let $\mathbb{B}$ be the Baire field of G. Then we have $\mathbb{B}_u(\tau) = \mathbb{B} \cap G_1$ and $\mathbb{B}_u(\tau') = \mathcal{B}(\mathbb{B} \cap G_0 \times \mathbb{B}_{\mathbb{Z}})$, $\mathbb{B}_{\mathbb{Z}}$ being the family of all subsets of $\mathbb{Z}$. Evidently we have $\mathbb{B}_u(\tau) \subset \mathbb{B}_u(\tau')$.

Let μ be the Haar measure on G. The Haar measure ν on G_1 in the topology τ is the trace of μ on G_1, while ν' in the topology τ' is the product of the trace of μ on G_0 and the Haar measure on $\mathbb{Z}$. Then, ν can not be extended to a right-invariant measure on $\mathbb{B}_u(\tau')$, while the restriction of ν' on $\mathbb{B}_u(\tau)$ is not σ-finite (it attains only 0 and ∞ on $\mathbb{B}_u(\tau)$).

This is an example in which a group G_1 becomes a thick group in two different ways, the corresponding Haar measures being essentially different.

§4. Weil topology

In this section we shall discuss the converse of the results in §2 and §3.

Definition 4.1. Let $(G, \mathcal{B})$ be a measurable group. A measure μ on $(G, \mathcal{B})$ is said to be separative, if

(4.1) $x \neq e \Longrightarrow {}^{\exists}E \in \mathcal{B},\ \mu(E)>0,\ \ \mu(E \cap Ex)=0.$

The Haar measure on a thick group is separative, because the topology is Hausdorff and a neighbourhood of e has a positive measure.

Theorem 4.1. Suppose that a right-invariant, separative measure μ exists on a measurable group $(G,\mathcal{B})$. Put

(4.2) $\mathfrak{A}=\{E \in \mathcal{B};\ 0<\mu(E)<\infty\}$, and

(4.3) $W_{E,\varepsilon}=\{x \in G;\ \mu(E \Delta Ex)<\varepsilon\}$,

then $\{W_{E,\varepsilon}\}_{E \in \mathfrak{A}, \varepsilon>0}$ becomes a fundamental system of neighbourhoods of e in some topology τ. G becomes a thick group in the topology τ, $\mathcal{B}$ contains $\mathbb{B}_u(\tau)$ and the restriction of μ on $\mathbb{B}_u(\tau)$ becomes the right Haar measure on $(G,\mathbb{B}_u(\tau))$.

This theorem claims that a right-invariant separative measure does not exist besides a possible extension of the right Haar measure.

Definition 4.2. The topology τ defined in Th. 4.1 is called the Weil topology of μ (or of $(G,\mathcal{B})$).

Proof of Th. 4.1. First Step. For $E \in \mathfrak{A}$, put

(4.4) $d_E(x,y)=\mu(Ex \Delta Ey)$,

then $d_E(x,y)$ is a right-invariant pseudo-metric. Namely it satisfies

(4.5) $d_E(x,x)=0$, $d_E(x,y)=d_E(y,x)$,

$d_E(x,z)\leqq d_E(x,y)+d_E(y,z)$ and

(4.6) $d_E(xz,yz)=d_E(x,y)$.

The family $\{d_E\}_{E\in\mathfrak{A}}$ defines a uniform structure on G, in which $\{W_{E,\varepsilon}\}_{E\in\mathfrak{A},\varepsilon>0}$ is a fundamental system of neighbourhoods of e. The multiplication is continuous because $d_E(xy,x_0y_0)\leqq d_E(xy,x_0y)+d_E(x_0y,x_0y_0)=d_E(x,x_0)+d_{Ex_0}(y,y_0)$. The inverse operation is continuous because $d_E(x^{-1},x_0^{-1})=d_E(e,x_0^{-1}x)=d_{Ex_0^{-1}}(x_0,x)$.

Thus G is a topological group in the Weil topology τ, and $\{d_E\}_{E\in\mathfrak{A}}$ gives the right uniform structure of τ. τ is Hausdorff because μ is separative.

Second Step. For $E\in\mathfrak{A}$ and $\varepsilon>0$, the set $W_{E,\varepsilon}$ belongs to $\mathcal{B}$, because $\mu(E\Delta Ex)$ is a $\mathcal{B}$-measurable function of x. (For any $\mathcal{B}\times\mathcal{B}$ -measurable set A, Fubini's theorem assures that $\mu(A(y))$ is a $\mathcal{B}$-measurable function of y, where $A(y)=\{x;\ (x,y)\in A\}$. On the other hand we have $E\Delta Ey=(\Phi^{-1}(E)\Delta p_1^{-1}(E))(y)$ where Φ: $(x,y)\to xy^{-1}$ and p_1: $(x,y)\to x$).

Third Step. We have $\mu(W_{E,\varepsilon})>0$ for any $E\in\mathfrak{A}$ and $\varepsilon>0$.

Consider $A=\Phi^{-1}(E)\cap p_2^{-1}(E)$ where p_2: $(x,y)\to y$, then

$$(\mu\times\mu)(A)=\int_G \mu(A(y))d\mu(y)=\int_E \mu(Ey)d\mu(y)=\mu(E)^2<\infty.$$

Since $\mathcal{B}\times\mathcal{B}$ is generated by $\{B\times B',\ B\in\mathcal{B},\ B'\in\mathcal{B}\}$, the set A can be approximated by a finite union of such measurable rectangles, i.e.

$$(4.7)\qquad \exists A'=\bigcup_{k=1}^{n}(B_k\times B_k'),\quad B_k\in\mathcal{B}\ ,\quad B_k'\in\mathcal{B}$$

$$(\mu\times\mu)(A\Delta A')<\varepsilon\mu(E)/2.$$

Here, we can assume that $\{B_k'\}$ is mutually disjoint. The left hand side of (4.7) is equal to $\int_E \mu(Ey\Delta A'(y))d\mu(y)$, so that putting $C=\{y;\mu(Ey\Delta A'(y))<\varepsilon/2\}$, we have $\mu(C)>0$. Put $C\cap B_k'=C_k$, then we have $\mu(C_k)>0$ for some k. On the other hand, $A'(y)=B_k$ for $y\in B_k'$, so that $C_k=\{y\in B_k';\ \mu(Ey\Delta B_k)<\varepsilon/2\}$. Let y_k be a fixed point of C_k, then $y\in C_k$ implies $d_E(y,y_k)<\varepsilon$, hence $yy_k^{-1}\in W_{E,\varepsilon}$, namely $C_ky_k^{-1}\subset W_{E,\varepsilon}$. Since C_k has a positive measure, this derives $\mu(W_{E,\varepsilon})>0$.

By a similar but rather complicated discussion, we can show $\mu(\bigcap_{n=1}^{N} W_{E_n,\varepsilon})>0$ for $E_n\in\mathfrak{M}$ $(n=1,2,\ldots,N)$ and $\varepsilon>0$. Thus every neighbourhood of e has a positive measure.

Fourth Step. If $E\in\mathfrak{M}$ and $\mu(E^{-1})<\infty$, then $\varepsilon<2\mu(E)$ implies $\mu(W_{E,\varepsilon})<\infty$.

By Fubini's theorem we have

$$\int_G \mu(Ey\cap E)d\mu(y)=(\mu\times\mu)\{\Phi^{-1}(E)\cap p_1^{-1}(E)\}$$
$$=\int_E \mu(E^{-1}x)d\mu(x)=\mu(E)\mu(E^{-1})<\infty.$$

Therefore we have

$$(4.8)\qquad \forall\alpha>0,\quad \mu(\{y;\ \mu(Ey\cap E)>\alpha\})<\infty.$$

Since $\mu(Ey\cap E)>\alpha$ is equivalent to $\mu(E\Delta Ey)<2(\mu(E)-\alpha)$, putting

$\varepsilon=2(\mu(E)-\alpha)$ (or $\alpha=\mu(E)-\varepsilon/2$), we have $\mu(W_{E,\varepsilon})<\infty$.

Fifth Step. G is locally precompact and σ-precompact in the Weil topology τ.

Take an $E\in\mathfrak{A}$ such that $\mu(E^{-1})<\infty$. Such E exists because μ is σ-finite. For this E and $\varepsilon<\mu(E)$, the set $W_{E,\varepsilon}$ is precompact as proved below.

Let U be an arbitrary symmetric neighbourhood of e, and V be a $\mathcal{B}$-measurable symmetric neighbourhood of e such that $V^2\subset U$. Let X be a maximal U-separated set contained in $W_{E,\varepsilon}$. (cf. Def. 2.3). Then $\{Ux\}_{x\in X}$ covers $W_{E,\varepsilon}$ and $\{Vx\}_{x\in X}$ is mutually disjoint. Supposing $V\subset W_{E,\varepsilon}$, every Vx is contained in $W_{E,2\varepsilon}$ so that $\#(X)\mu(V)\leq\mu(W_{E,2\varepsilon})<\infty$, hence $\#(X)<\infty$. This implies that $W_{E,\varepsilon}$ is precompact.

In a similar way, G can be covered by countable union of right translations of U. (This time, let X be a maximal U-separated set contained in G. Since μ is σ-finite, we have $^{\exists}E_n\in\mathcal{B}$, $\mu(E_n)<\infty$, $G=\bigcup_{n=1}^{\infty}E_n$. Put $X_n=\{x\in X;\ \mu(Vx\cap E_n)>0\}$, then X_n must be a countable set. Thus $X=\bigcup_{n=1}^{\infty}X_n$ is a countable set). If we take U as a precompact neighbourhood of e, then G turns out to be σ-precompact.

Sixth Step. We shall show $\mathbb{B}_u(\tau)\subset\mathcal{B}$.

Let f be a uniformly continuous function on G (in the Weil topology τ), then for any $\varepsilon>0$ there exists a $\mathcal{B}$-measurable neighbourhood $U(\varepsilon)$ of e such that $xy^{-1}\in U(\varepsilon)$ implies $|f(x)-f(y)|<\varepsilon$. We shall prove $f^{-1}((-\infty,a))\in\mathcal{B}$. Since $f^{-1}((-\infty,a-1/n))$ can be covered by countable union of right translations of $U(1/n)$, we have

(4.9) $\quad \exists B_n \in \mathcal{B}, \quad f^{-1}((-\infty,a-1/n)) \subset B_n \subset f^{-1}((-\infty,a))$.

So that we have $f^{-1}((-\infty,a)) = \bigcup_{n=1}^{\infty} B_n \in \mathcal{B}$.

Seventh Step. If $E \in \mathfrak{M}$ and $\mu(E^{-1}) < \infty$, then by Fourth Step we have $\mu(W_{E,\varepsilon}) < \infty$. Since every precompact measurable set B can be covered by finite union of right translations of $W_{E,\varepsilon}$, we have $\mu(B) < \infty$. Thus μ is locally finite. From Th. 3.4, G is a thick group and the restriction of μ on $\mathbb{B}_u(\tau)$ is the right Haar measure. (q.e.d.)

Theorem 4.2. Suppose that a right-invariant separative measure μ exists on a measurable group $(G,\mathcal{B})$. Then μ is a finite measure if and only if G is precompact in the Weil topology.

Proof is contained in the Fifth and Seventh Steps of Th. 4.1.

Corollary 1. Let G be a thick group. The Haar measure is finite if and only if G is precompact.

Corollary 2. Let G be a locally compact and σ-compact group. The Haar measure is finite if and only if G is compact.

§5. Case of a vector space

We shall apply the result in §4 to a vector space. Let X be a real vector space, X' be its algebraical dual space and R be a linear subspace of X'. Let $\mathcal{B}_R$ be the smallest σ-algebra on X in which every $f \in R$ is measurable.

Theorem 5.1. $(X, \mathcal{B}_R)$ is a measurable additive group.

Proof. We shall prove that Φ: $(x,y) \to x-y$ is a measurable map from $(X \times X, \mathcal{B}_R \times \mathcal{B}_R)$ to $(X,\mathcal{B}_R)$. Since $\mathcal{B}_R$ is generated by $\{f^{-1}(B);\ f \in R,\ B$ is a Borel subset of $\mathbb{R}^1\}$, it is sufficient to show that $\Phi^{-1}(f^{-1}(B)) \in \mathcal{B}_R \times \mathcal{B}_R$, in other words $f \circ \Phi$ is a measurable function with respect to $\mathcal{B}_R \times \mathcal{B}_R$. But since f is linear, we have $f \circ \Phi(x,y) = f(x) - f(y)$, and since the projections $(x,y) \to x$ and $(x,y) \to y$ are measurable, $f \circ \Phi$ turns out to be a measurable function. (q.e.d.)

Hereafter we shall assume that

$$(5.1) \qquad \forall x \neq 0, \quad \exists f \in R, \quad f(x) \neq 0.$$

Even if (5.1) is not satisfied, we can identify a measure on $(X, \mathcal{B}_R)$ with a measure on $X/R^{\perp}$ where $R^{\perp} = \{x \in X;\ f(x) = 0 \text{ for } \forall f \in R\}$. Thus the following discussions are kept valid by replacing X with $X/R^{\perp}$.

Theorem 5.2. An invariant measure μ on $(X, \mathcal{B}_R)$ is necessarily separative.

Proof. We shall prove that

$$(5.2) \qquad x \neq 0 \Rightarrow \exists E \in \mathcal{B}_R, \quad \mu(E) > 0, \quad E \cap (E+x) = \phi.$$

From (5.1) there exists $f \in R$ such that $f(x) \neq 0$. We can suppose that $f(x) = 1$. Then putting $E = f^{-1}([0,1))$, we see that the family $\{E+kx\}$ $(k = 0, \pm 1, \pm 2, \cdots)$ becomes a partition of X. This E satisfies (5.2). (q.e.d.)

Theorem 5.3. If X is infinite dimensional, there exists no invariant measure on $(X, \mathcal{B}_R)$.

Proof. We shall show that if an invariant measure μ exists on $(X, \mathcal{B}_R)$, then X is finite dimensional.

Since μ is separative, Th. 4.1 assures that X is a locally precompact additive group in the Weil topology. We shall define C-topology (=the convexified Weil topology) as follows: Let $\mathcal{U}=\{U\}$ be the family of all symmetric neighbourhoods of 0 in the Weil topology. Let $C(U)$ be the convex closure of U, i.e.

(5.3) $$C(U)=\{\lambda_1x_1+\lambda_2x_2+\cdots+\lambda_nx_n;\ n=1,2,\cdots,\quad \lambda_i>0,\ \sum_{i=1}^{n}\lambda_i=1\}.$$

For each $U\in\mathcal{U}$, $C(U)$ is convex and symmetric. If we can show that $C(U)$ is absorbant, i.e.

(5.4) $$\forall x\neq 0,\quad \exists\lambda_0>0,\quad \lambda_0x\in C(U),$$

then $\{C(U)\}_{U\in\mathcal{U}}$ becomes a fundamental system of neighbourhoods of 0 in a locally convex topology on X. This topology is called C-topology.

We shall prove (5.4). Since X is a σ-precompact additive group in the Weil topology, any U-separated set should be countable. However, the set $\{\lambda x;\ \lambda\in\mathbb{R}^1\}$ being non-countable, it is not U-separated so that we have $\exists\lambda_1\neq\lambda_2$, $\lambda_1x-\lambda_2x\in U$, hence we have $\lambda_0x\in U\subset C(U)$ for $\lambda_0=\lambda_1-\lambda_2\neq 0$.

X is Hausdorff in C-topology, because for any $f\in R$, $f^{-1}((-1,1))$ is a convex neighbourhood of 0 in the Weil topology, (cf. Proof of Th. 5.2), hence also in C-topology.

Since X is locally precompact in the Weil topology, some $U\in\mathcal{U}$ is precompact in the Weil topology. Since C-topology is weaker than the Weil topology, U is precompact also in C-topology. Thus $C(U)$ is precompact in C-topology as proved in the following lemma 1, therefore X is locally precompact in C-topology. So that lemma 2 assures that X is finite dimensional.

<u>Lemma 1</u>. Let X be a locally convex vector space. If A is a precompact subset of X, then the convex closure $C(A)$ is also precompact.

<u>Proof.</u> For any neighbourhood U of 0, we shall prove that

$C(A)$ is covered by a finite union of translations of U. Let V be a convex neighbourhood of 0 such that $V+V\subset U$. Since A is precompact, we have $A\subset \bigcup_{i=1}^{n}(V+x_i)$ for some $\{x_i\}_{i=1}^{n}$. Let B be the convex closure of $\{x_1,x_2,\cdots,x_n\}$, then B is evidently compact so that we have $B\subset \bigcup_{j=1}^{m}(V+y_j)$ for some $\{y_j\}_{j=1}^{m}$. Since $A\subset V+B$ and $V+B$ is convex, we have $C(A)\subset V+B\subset \bigcup_{j=1}^{m}(V+V+y_j)\subset \bigcup_{j=1}^{m}(U+y_j)$. (q.e.d.)

Lemma 2. Let X be a topological vector space. If X is locally precompact, then X is finite dimensional.

Proof. Let U be a precompact neighbourhood of 0. Since $\frac{1}{2}U$ is a neighbourhood of 0, there exists some $\{x_i\}_{i=1}^{n}$ such that

$$(5.5)\qquad U\subset \bigcup_{i=1}^{n}\left(\tfrac{1}{2}U+x_i\right).$$

Let R be the linear subspace spanned by $\{x_i\}_{i=1}^{n}$, then we have $U\subset \frac{1}{2}U+R$. Therefore $U\subset \frac{1}{2}(\frac{1}{2}U+R)+R=\frac{1}{4}U+R$. Repeating this, we get $U\subset \frac{1}{2^k}U+R$ for any $k>0$.

Take an arbitrary element $x\in U$, then for some $u_k\in U$ and $y_k\in R$ we have

$$(5.6)\qquad x = \frac{1}{2^k}u_k+y_k.$$

Let R_1 be the linear space spanned by x and R. Since the topology on a finite dimensional vector space is unique (lemma 3), $U\cap R_1$ is precompact in the Euclidean topology on R_1, so that it is bounded in the Euclidean topology, thus we get $\lim_{k\to\infty}\frac{1}{2^k}u_k=0$. This means $x=\lim_{k\to\infty}y_k\in R$, hence $U\subset R$. Since U is absorbant, we have $X=R$, thus X is finite dimensional.

(q.e.d.)

Lemma 3. An n-dimensional topological vector space X is necessarily isomorphic to $\mathbb{R}^n$.

Proof. Consider a basis $\{e_1, e_2, \cdots, e_n\}$ of X. Then the map Φ: $\mathbb{R}^n \ni (\lambda_1, \lambda_2, \cdots, \lambda_n) \to \lambda_1 e_1 + \lambda_2 e_2 + \cdots + \lambda_n e_n \in X$ is evidently an algebraic isomorphism. Φ is continuous, because the scalar multiplication and the addition are continuous on X.

Let B be the closed unit ball and S be the unit sphere in $\mathbb{R}^n$. If $\Phi(B)$ contains some neighbourhood of 0 in X, then Φ turns out to be a homeomorphism. Since $\Phi(S)$ is compact, there exists a neighbourhood U of 0 such that $U \cap \Phi(S) = \phi$. Then, a positive number λ_0 and a neighbourhood V of 0 exist such that $|\lambda| \leq \lambda_0$ implies $\lambda V \subset U$, hence $\lambda_0 V \subset \bigcap_{t \geq 1} (tU)$. This implies $\lambda_0 \Phi^{-1}(V) \cap \bigcup_{t \geq 1} (tS) = \phi$ or equivalently $\lambda_0 \Phi^{-1}(V) \subset B$, hence $\lambda_0 V \subset \Phi(B)$. (q.e.d.)

Corollary 1 (of Th. 5.3). If X is infinite dimensional, there exists no quasi-invariant measure on $(X, \mathcal{B}_R)$.

Proof is obtained by Th. 5.3 and Th. 1.4.

Corollary 2. If X is infinite dimensional and if $\mathcal{B}$ is a σ-algebra which contains $\mathcal{B}_R$ for some $R \subset X'$, then there exists neither invariant nor quasi-invariant measure on $(X, \mathcal{B})$.

Proof. If μ is a σ-finite and invariant (or quasi-invariant) measure on $(X, \mathcal{B})$, then there exists a finite measure ν which is equivalent with μ in the sense of absolute continuity. Then the restriction of ν on $\mathcal{B}_R$ is finite and quasi-invariant on $(X, \mathcal{B}_R)$. This is a contradiction to Corollary 1.

(q.e.d.)

Remark. Though μ is σ-finite on $\mathcal{B}$, it may not be σ-finite on $\mathcal{B}_R$, so the restriction of μ on $\mathcal{B}_R$ is not useful for the proof.

Especially, considering the case $R=X^*$ (=the topological dual space of X) we have

Corollary 3. On a infinite dimensional locally convex space, there exists neither invariant nor quasi-invariant Borel measure.

The same holds even if we replace Borel measure by Baire measure.

§6. Quasi-invariance and ergodicity

Let $(X,\mathcal{B})$ be a measurable space and φ be a $\mathcal{B}$-automorphism. This means that φ is a bijection from X onto X such that $\mathcal{B}=\varphi^{-1}(\mathcal{B})$. The image measure $\varphi\circ\mu$ is defined by

(6.1) $\varphi\circ\mu(E)=\mu(\varphi^{-1}(E))$ for $^{\forall}E\in\mathcal{B}$.

Definition 6.1. Let G be a group of $\mathcal{B}$-automorphisms. A measure μ on $(X,\mathcal{B})$ is said to be G-invariant, if $\varphi\circ\mu=\mu$ for every $\varphi\in G$. μ is said to be G-quasi-invariant if $\varphi\circ\mu\sim\mu$ for every $\varphi\in G$.

Every G-invariant measure is G-quasi-invariant. However, even if μ is G-quasi-invariant, μ does not always have an equivalent G-invariant measure. (In this lecture, a measure is always assumed to be σ-finite).

Theorem 6.1. For a G-quasi-invariant measure μ, the following five statements are equivalent.

(E1) If $\mu\sim\mu_1+\mu_2$ where both μ_1 and μ_2 are G-quasi-invariant and $\not\equiv 0$, then $\mu\sim\mu_1$ and $\mu\sim\mu_2$.

(E2) If $\mu_1\leq\mu$ and μ_1 is G-quasi-invariant, then $\mu_1\sim\mu$ or $\mu_1\equiv 0$.

(E3) If $E\in\mathcal{B}$ and $\mu(E)>0$, then $^{\exists}\{\varphi_k\}_{k=1,2,\ldots}\subset G$, $\mu((\bigcup_{k=1}^{\infty}\varphi_k(E))^c)=0$.

(E4) If $E\in\mathcal{B}$ and $\mu(E\Delta\varphi(E))=0$ for every $\varphi\in G$, then $\mu(E)=0$ or $\mu(E^c)=0$.

(E5) If $f(x)$ is $\mathcal{B}$-measurable and $^{\forall\prime}x$, $f(x)=f(\varphi(x))$ for every $\varphi\in G$, then $f(x)$ is constant almost everywhere.

Theorem 6.1'. For a G-invariant measure μ, the following two statements are equivalent. They are equivalent also with (E1)~(E5).

(E6) If $\mu=\mu_1+\mu_2$ where both μ_1 and μ_2 are G-invariant, then $\mu_1=a\mu$ and $\mu_2=(1-a)\mu$ for some constant $0\leqq a\leqq 1$.

(E7) If $\mu_1\leqq\mu$ and μ_1 is G-invariant, then $\mu_1=a\mu$ for some constant $0\leqq a\leqq 1$.

Definition 6.2. A G-quasi-invariant (resp. G-invariant) measure μ is said to be G-ergodic, if it satisfies any of the conditions (E1)~(E5). (resp. (E1)~(E7)).

Proof. (E1) and (E2) are equivalent, because $\mu\sim\mu_1+\mu_2$ implies $\mu_1\lesssim\mu$ and $\mu_2\lesssim\mu$, and $\mu_1\lesssim\mu$ implies $\mu\sim\mu_1+\mu$.

Assuming (E2), we have (E3) as follows. Supposing $\mu(E)>0$, put

$$(6.2)\qquad \mu_1(B)=\inf_F\{\mu(B\cap F^c);\ ^\exists\{\varphi_k\}_{k=1,2,\dots}\subset G,\ F=\bigcup_{k=1}^{\infty}\varphi_k(E)\}.$$

The infimum is attained by some F, namely $^\forall B\in\mathcal{B}$, $^\exists F=\bigcup_{k=1}^{\infty}\varphi_k(E)$, $\mu_1(B)=\mu(B\cap F^c)$. The set F can be taken common for countable number of sets B, thus μ_1 is a σ-additive measure. From the definition (6.2), μ_1 is G-quasi-invariant and $\mu_1\lesssim\mu$. Since $\mu_1(E)=0$, we have $\mu_1\not\sim\mu$, hence from (E2) we have $\mu_1\equiv 0$. This means $^\exists F$, $\mu(F^c)=0$.

Assuming (E3), we have (E4) as follows. If $\mu(E\Delta\varphi(E))=0$ for $^\forall\varphi\in G$, then $\mu(E\Delta\bigcup_{k=1}^{\infty}\varphi_k(E))=0$ so that (E3) implies $\mu(E)=0$ or $\mu(E^c)=0$.

Assuming (E4), we have (E5). If $^\forall\varphi\in G$, $^{\forall'}x$, $f(x)=f(\varphi(x))$, then $E_\alpha=f^{-1}((\alpha,\infty))$ satisfies the assumption of (E4), so that $\mu(E_\alpha)=0$ or $\mu(E_\alpha^c)=0$. Thus the image measure $f\circ\mu$ is a Dirac measure on $\mathbb{R}^1$. This means that $f(x)$ is constant almost everywhere.

Conversely (E5) implies (E4), because $\mu(E\Delta\varphi(E))=0$ means that the indicator function $C_E(x)$ (=1 for $x\in E$, =0 for $x\notin E$) satisfies the assumption of (E5), so that $\forall' x$, $C_E(x)=0$ or $\forall' x$, $C_E(x)=1$.

(E4) implies (E2). $\mu_1 \lesssim \mu$ means that

$$(6.3) \qquad \exists f(x) \geqq 0, \; d\mu_1 = f d\mu.$$

Put $E=f^{-1}(0)$, then G-quasi-invariance of μ_1 implies $\mu(E\Delta\varphi(E))=0$, thus $\mu(E)=0$ or $\mu(E^c)=0$. The former means $\mu_1 \sim \mu$ and the latter means $\mu_1 \equiv 0$.

Evidently (E7) implies (E6). To show that (E6) implies (E7), we must prove that $\mu_1 \leqq \mu$ implies $\mu=\mu_1+\mu_2$ for some μ_2. If μ is finite, we can put $\mu_2=\mu-\mu_1$. If μ is σ-finite, put $\mu_2(B)=\sum_{n=1}^{\infty}\{\mu(E_n)-\mu_1(E_n)\}$ where $B=\bigcup_{n=1}^{\infty} E_n$, $\{E_n\}$ is disjoint and $\mu(E_n)<\infty$. $\mu_2(B)$ is independent of the choice of $\{E_n\}$ and becomes a σ-additive measure such that $\mu=\mu_1+\mu_2$.

(E7) is equivalent with (E5). First, assume (E5). If $\mu_1 \leqq \mu$ where both μ and μ_1 are G-invariant, then the density function $d\mu_1/d\mu$ satisfies the assumption of (E5), so it is constant almost everywhere. This means $\mu_1=a\mu$ for some constant a. Next, assume (E7). We shall prove (E5) assuming $1> f(x)>0$. Put $d\mu_1=fd\mu$, then μ_1 is G-invariant so that $\mu_1=a\mu$. This means $f(x)=a$ almost everywhere. (q.e.d.)

<u>Remark 1</u>. Let G_1 and G_2 be groups of $\mathcal{B}$-automorphisms such that $G_1 \subset G_2$. Then, every G_2-(quasi)-invariant measure is G_1-(quasi)-invariant. For a G_2-quasi-invariant measure, if it is

G_1-ergodic, then it is G_2-ergodic also.

Remark 2. Let G be a group of $\mathcal{B}$-automorphisms and φ be another $\mathcal{B}$-automorphism. If μ is G-(quasi)-invariant, then $\varphi\circ\mu$ is $\varphi G\varphi^{-1}$-(quasi)-invariant. If μ is G-ergodic, then $\varphi\circ\mu$ is $\varphi G\varphi^{-1}$-ergodic.

Remark 3. Let μ,ν be G-ergodic measures. Then we have $\mu\sim\nu$ or $\mu\perp\nu$, because $\mu(E)=0$ and $\nu(E)>0$ imply $\mu(F)=0$ and $\nu(F^c)=0$ for some $F=\bigcup_{k=1}^{\infty}\varphi_k(E)$.

§7. Absolute continuity of projective limit measures

Let $\mathcal{B}_n$ be an increasing sequence of σ-algebras of subsets of X. Let μ and μ' be probability measures on $\mathcal{B}=\mathcal{B}(\bigcup_{n=1}^{\infty}\mathcal{B}_n)$, and denote the restriction of μ (resp. μ') on $\mathcal{B}_n$ with μ_n (resp. μ_n').

Theorem 7.1. We have $\mu'\lesssim\mu$, if and only if $\forall n$, $\mu_n'\lesssim\mu_n$ and $\{(d\mu_n'/d\mu_n)^{1/p}\}$ converges in $L^p(\mu)$ where $1\leqq p<\infty$.

Corollary (Case for $p=2$). We have $\mu'\lesssim\mu$, if and only if $\forall n$, $\mu_n'\lesssim\mu_n$ and

$$\int_X\sqrt{\frac{d\mu_n'}{d\mu_n}(x)\frac{d\mu_m'}{d\mu_m}(x)}\,d\mu(x)\xrightarrow[(n,m\to\infty)]{}1. \tag{7.1}$$

Proof (of Theorem). The condition $\forall n$, $\mu_n'\lesssim\mu_n$ is evidently necessary. Assume this condition and put $f_n=d\mu_n'/d\mu_n$. If $f_n^{1/p}$ converges to g in $L^p(\mu)$, then $f_n^{1/p}\varphi$ converges to $g\varphi$ for a bounded function φ, so that $\int_X f_n|\varphi|^p d\mu$ converges to $\int_X|g\varphi|^p d\mu$. Taking φ as the indicator function of $E\in\mathcal{B}_m$, we have $\int_X f_n c_E d\mu=\mu_n'(E)=\mu'(E)$ for $n\geqq m$, so that $\int_E|g|^p d\mu=\mu'(E)$. This means that $d\mu'=|g|^p d\mu$, hence $\mu'\lesssim\mu$.

Conversely, assume that $\mu'\lesssim\mu$. Since $|a^{1/p}-b^{1/p}|^p\leqq|a-b|$

for a≧0, b≧0 and p≧1, if $\{f_n\}$ converges in $L^1(\mu)$, then $\{f_n^{1/p}\}$ converges in $L^p(\mu)$ for p>1.

Now, we shall prove $\|f_n-f\|_1\to 0$ where $f=d\mu'/d\mu$. Since $\mathcal{B}=\mathcal{B}(\bigcup_{n=1}^{\infty}\mathcal{B}_n)$, f can be written as $f=\lim_{n\to\infty} g_n$ in $L^1(\mu)$, where g_n is $\mathcal{B}_n$-measurable. So it is sufficient to prove $\|f_n-g_n\|_1\leqq \|f-g_n\|_1$.

<u>Lemma 1</u>. For every $\mathcal{B}_n$-measurable summable function h(x), we have

$$(7.2) \qquad \|f_n-h\|_1 \leqq \|f-h\|_1$$

where $f_n=d\mu_n'/d\mu_n$ and $f=d\mu'/d\mu$.

<u>Proof</u>. Put $E=\{x;f_n(x)\geqq h(x)\}$, then $E\in\mathcal{B}_n$, so that $\int_E f_n(x)d\mu(x) =\int_E f(x)d\mu(x)$. Therefore

$$\begin{aligned}\|f_n-h\|_1 &= \int_E\{f_n(x)-h(x)\}d\mu(x)-\int_{E^c}\{f_n(x)-h(x)\}d\mu(x)\\ &= \int_E\{f(x)-h(x)\}d\mu(x)-\int_{E^c}\{f(x)-h(x)\}d\mu(x)\\ &\leqq \int_X|f(x)-h(x)|d\mu(x) = \|f-h\|_1. \qquad \text{(q.e.d.)}\end{aligned}$$

<u>Theorem 7.2</u>. If $\mu'\leqq\mu$, then $d\mu_n'/d\mu_n$ converges to $d\mu'/d\mu$ almost everywhere in μ.

<u>Proof</u>. From Th.7.1, $f_n=d\mu_n'/d\mu_n$ converges to $f=d\mu'/d\mu$ in $L^1(\mu)$. Therefore it is sufficient to prove that $f_n(x)$ converges (to some function) almost everywhere, namely that $g_n(x) =\sup_{m>n}|f_n(x)-f_m(x)|$ converges to 0 in probability.

Put $A_{n,\varepsilon}=\{x;g_n(x)>\varepsilon\}$ and consider the partition $A_{n,\varepsilon}= \bigcup_{m=n+1}^{\infty} A_{n,m,\varepsilon}$ where $A_{n,m,\varepsilon}=\{x;|f_n(x)-f_m(x)|>\varepsilon,\ |f_n(x)-f_k(x)|\leqq\varepsilon$ for $n\leqq k<m\}$. Since $A_{n,m,\varepsilon}\in\mathcal{B}_m$, we get (similarly as the proof of Lemma 1)

$$\int_{A_{n,m,\varepsilon}} |f(x)-f_n(x)|d\mu(x) \geqq \int_{A_{n,m,\varepsilon}} |f_m(x)-f_n(x)|d\mu(x)$$
$$>\varepsilon\mu(A_{n,m,\varepsilon})$$

so that summing up with respect to m, we have $\|f-f_n\|_1>\varepsilon\mu(A_{n,\varepsilon})$, thus $\lim_{n\to\infty} \mu(A_{n,\varepsilon})=0$. This means that g_n converges to 0 in probability. (q.e.d.)

Theorem 7.3. Suppose that $\forall n$, $\mu_n' \lesssim \mu_n$, then $d\mu_n'/d\mu_n$ converges almost everywhere in μ (even if $\mu' \not\lesssim \mu$). The limit function $f(x)$ satisfies $d\mu_c'=fd\mu$ where μ_c' is the continuous part of Radon-Nikodym's decomposition of μ' with respect to μ. We have $\mu' \lesssim \mu$ if and only if $\|f\|_1=1$.

Proof. Since $\mu \lesssim \mu+\mu'$ and $d\mu_n/d(\mu_n+\mu_n')=(1+f_n(x))^{-1}$, from Th. 7.2 $(1+f_n(x))^{-1}$ converges almost everywhere (in $\mu+\mu'$) to a function $g(x)$ such that $d\mu=gd(\mu+\mu')$. Then, putting $d\mu_c'=fd\mu$, we have $d\mu=g(1+f)d\mu+gd\mu_s'$. From the uniqueness of Radon-Nikodym's decomposition we have $g(1+f)=1$ almost everywhere in μ and $gd\mu_s'=0$. The former means $g=(1+f)^{-1}$, so that $f_n(x)$ converges to $f(x)$ almost everywhere in μ. Finally, $\mu' \lesssim \mu$ is equivalent with $\mu_s'=0$ (μ_s' is the singular part of Radon-Nikodym's decomposition), hence with $\mu_c'(X)=1$, hence with $\|f\|_1=1$. (q.e.d.)

Let $\mathcal{L}=\{L\}$ be a directed set, and $\{\mathcal{B}_L\}$ be an increasing family of σ-algebras parametrized by $L\in\mathcal{L}$. Let μ and μ' be probability measures on $\mathcal{B}=\mathcal{B}(\bigcup_{L\in\mathcal{L}}\mathcal{B}_L)$, and denote the restriction of μ (resp. μ') on $\mathcal{B}_L$ with μ_L (resp. μ_L'). Similarly as Th.7.1 and Th.7.3, we get the following theorems.

Theorem 7.1'. We have $\mu' \lesssim \mu$, if and only if $\forall L\in\mathcal{L}$, $\mu_L' \lesssim \mu_L$ and $\{(d\mu_L'/d\mu_L)^{1/p}\}$ converges in $L^p(\mu)$ where $1 \leqq p<\infty$.

Theorem 7.3'. Suppose that $\forall L\in\mathcal{L}$, $\mu'_L\lesssim\mu_L$, then for every increasing sequence $M=\{L_n\}\subset\mathcal{L}$, $d\mu'_{L_n}/d\mu_{L_n}$ converges almost everywhere in μ. Denote the limit function with $f_M(x)$. We can find $M_0=\{L_{n0}\}$ such that $f_M(x)$ is independent of M if M satisfies $\forall n, \exists m, L_m\geq L_{n0}$. This function $f(x)$ $(=f_M(x)$ for such $M)$ satisfies $d\mu'_c=fd\mu$. We have $\mu'\lesssim\mu$ if and only if $\|f\|_1=1$.

The existence of M_0 in Th.7.3' is derived from the fact that $d\mu'_c/d\mu$ is $\mathcal{B}$-measurable so that $\exists M_0$, $\mathcal{B}(\bigcup_{n=1}^{\infty}\mathcal{B}_{L_{n0}})$-measurable.

Example. Let $(X,\mathcal{B}_1)$ be a measurable space and μ_1,μ'_1 be probability measures on $\mathcal{B}_1$. Denote the infinite product of μ_1 (resp. μ'_1) with μ (resp. μ'). We have $\mu\perp\mu'$ whenever $\mu_1\neq\mu'_1$.

First we shall prove $\mu\not\sim\mu'$. If $\mu_1\not\sim\mu'_1$, then evidently $\mu\not\sim\mu'$. If $d\mu'_1=f_1d\mu_1$, then since μ and μ' are product measures we have $d\mu'_n/d\mu_n=f_n(x_1,x_2,\cdots,x_n)=\prod_{k=1}^{n}f_1(x_k)$. Therefore we get $\langle\sqrt{f_n},\sqrt{f_m}\rangle=\langle\sqrt{f_1},1\rangle^{m-n}$. From Corollary of Th.7.1, $\mu\sim\mu'$ implies $\langle\sqrt{f_1},1\rangle=1$, hence $f_1=1$. This means $\mu_1=\mu'_1$.

Thus we have seen $\mu\not\sim\mu'$. From Remark 3 of §6, if μ and μ' are G-ergodic, then we see $\mu\perp\mu'$.

Lemma 2. Let μ be the infinite product of an identical probability measure μ_1. Let G be the group of all permutations of natural numbers. G can be considered as a transformation group of X^{∞}: $G\ni\sigma$, $(x_n)\to(x_{\sigma(n)})$. Then, μ is G-ergodic.

Proof. From the definition of the product σ-algebra, we have $\mathcal{B}=\mathcal{B}(\bigcup_{n=1}^{\infty}\mathcal{B}_n)$ where $\mathcal{B}_n$ is the finite product of $\mathcal{B}_1$. So every $E\in\mathcal{B}$ satisfies $\forall\varepsilon>0$, $\exists n$, $\exists E_n\in\mathcal{B}_n$, $\mu(E\Delta E_n)<\varepsilon$.

Now, suppose that $\mu(E\Delta\sigma(E))=0$ for $\forall\sigma\in G$. Since μ is

G-invariant, we have $\mu(E_n \Delta \sigma(E_n))<2\varepsilon$. Take such σ that $\{1, 2,\cdots,n\}$ is disjoint with $\{\sigma(1),\sigma(2),\cdots,\sigma(n)\}$. Then, we have $\mu(E_n \Delta \sigma(E_n))=2\mu(E_n)(1-\mu(E_n))$, hence $\mu(E_n)(1-\mu(E_n))<\varepsilon$. Combining this with $\mu(E \Delta E_n)<\varepsilon$, we get $\mu(E)(1-\mu(E))<2\varepsilon$. Since $\varepsilon>0$ is arbitrary, we must have $\mu(E)(1-\mu(E))=0$, hence $\mu(E)=0$ or 1. This means that μ is G-ergodic. (q.e.d.)

§8. Gaussian measures

One-dimensional gaussian measure g_c (with variance c^2) is defined by

$$(8.1) \qquad dg_c(t) = \frac{1}{\sqrt{2\pi}c} \exp(-\frac{t^2}{2c^2})dt.$$

Here dt means the Lebesgue measure on $\mathbb{R}^1$. We shall include Dirac measure δ ($\delta(A)=1$ if $A\ni 0$, $=0$ if $A\not\ni 0$) as a gaussian measure with variance 0, so hereafter we shall assume $c\geqq 0$.

Let E be a real vector space, E^a be its algebraical dual space, and $\mathcal{B}_E$ be the smallest σ-algebra in which $E^a \ni x \to x(\xi)$ is measurable for $\forall \xi \in E$.

Theorem 8.1. For a measure μ on $(E^a,\mathcal{B}_E)$, the following two statements are equivalent.

1) The characteristic function χ of μ is written in the form of

$$(8.2) \qquad \chi(\xi) = \exp(-\tfrac{1}{2}(\xi,\xi))$$

for some inner product $(\ ,\ )$ on E.

2) For every $\xi \in E$, the distribution of $x(\xi)$ becomes a one-dimensional gaussian measure.

Definition 8.1. A measure μ is called a gaussian measure on

$(E^a,\mathcal{B}_E)$, if it satisfies 1) or 2) in Th.8.1. If a subspace X of E^a is thick in μ, the trace of μ on X is called a gaussian measure on $(X,\mathcal{B}_E \cap X)$.

Proof. First assume 1). Let R be a finite dimensional subspace of E and $\{e_k\}_{k=1}^n$ be a base of R such that $\{e_k\}_{k=1}^m$ is orthonormal and $(e_k,e_k)=0$ for $m<k\leq n$. Then we have

$$(8.3) \qquad \chi(\sum_{k=1}^n \alpha_k e_k) = \exp(-\frac{1}{2}\sum_{k=1}^m \alpha_k^2)$$
$$= \frac{1}{(2\pi)^{m/2}}\int_{-\infty}^{\infty}\cdots\int_{-\infty}^{\infty}\exp[i\sum_{k=1}^m \alpha_k t_k - \frac{1}{2}\sum_{k=1}^m t_k^2]dt_1 dt_2\cdots dt_m.$$

This implies that $\chi(\xi)$ is positive definite, so that χ corresponds to a measure on $(E^a,\mathcal{B}_E)$. (c.f. §16 of Part A).

Especially for a given $\xi\in E$, putting $e_1=\frac{\xi}{\|\xi\|}$ we see

$$\chi(s\xi) = \frac{1}{\sqrt{2\pi}\|\xi\|}\int_{-\infty}^{\infty}\exp(ist - \frac{1}{2}\frac{t^2}{\|\xi\|^2})dt.$$

Combining this with $\chi(s\xi)=\int_{E^a}\exp(isx(\xi))d\mu(x)$, we see that the distribution of $x(\xi)$ becomes one-dimensional gaussian measure with variance $\|\xi\|^2$. (If $\|\xi\|=0$, then $x(\xi)$ follows Dirac measure).

Next, assume 2). If the distribution of $x(\xi)$ is one-dimensional gaussian measure with variance $v(\xi)$, then we have

$$\chi(\xi) = \frac{1}{\sqrt{2\pi v(\xi)}}\int_{-\infty}^{\infty}\exp(it - \frac{t^2}{2v(\xi)})dt = \exp(-\frac{1}{2}v(\xi)).$$

On the other hand, we get

$$\int_{E^a} x(\xi)^2 d\mu(x) = \frac{1}{\sqrt{2\pi v(\xi)}}\int_{-\infty}^{\infty} t^2\exp(-\frac{t^2}{2v(\xi)})dt = v(\xi),$$

so that putting

$$(8.4) \qquad (\xi,\eta) = \int_{E^a} x(\xi)x(\eta)d\mu(x),$$

we get $v(\xi) = (\xi,\xi)$.. Evidently (8.4) defines an inner product on E. (q.e.d.)

We shall recall the following results. (cf. §19 of Part A).

Theorem 8.2. Let μ be a gaussian measure whose characteristic function is given by (8.2). Let E_1 be a subspace of E, $(\ ,\ )_1$ be an inner product on E_1, and E_1' ($\subset E^a$) be the set of all x such that the restriction of x on E_1 is continuous in $(\ ,\ )_1$.

1) If $(\ ,\ )$ is HS with respect to $(\ ,\ )_1$ on E_1, then E_1' is thick in μ.

2) Otherwise, the outer measure of E_1' is zero.

Corollary. If E is infinite dimensional (rigorously speaking, if E/M is infinite dimensional where $M=\{\xi\in E;\ (\xi,\xi)=0\}$), then the outer measure of E' (= topological dual space of E in $(\ ,\ )$) is zero.

Example. Let $E=\mathbb{R}_0^\infty$, and $(\ ,\ )$ be the usual inner product of (ℓ^2). Then the corresponding gaussian measure on $E^a=\mathbb{R}^\infty$ is the direct product of one-dimensional gaussian measures with variance 1. For a sequence $a=\{a_n\}$ of positive numbers, put

$$(8.5)\qquad H_a = \{x\in\mathbb{R}^\infty;\ \sum_{n=1}^{\infty} a_n^2 x_n^2 < \infty\}.$$

Then we have $\mu(H_a)=1$ for $a\in(\ell^2)$, and $\mu(H_a)=0$ for $a\notin(\ell^2)$. Note that $\bigcap_{a\in(\ell^2)} H_a=(\ell^\infty)$ but $\mu((\ell^\infty))=0$.

Remark. The above situation is seen more generally. Let E be a real vector space, and $(\ ,\ )$ be an inner product on E such that E/M (where $M=\{\xi;\ (\xi,\xi)=0\}$) is infinite dimensional. Let E_1 be a countable dimensional subspace of E and $(\ ,\ )_1$ be an inner product such that $(\ ,\ )$ is HS with respect to

$(\ , \ )_1$ on E_1. Then E_1' is thick in μ as stated in Th.8.2. However we have $E'=\bigcap_{E_1} E_1'$ and the outer measure of E' is zero.

'We shall prove $E'=\bigcap_{E_1} E_1'$. Since $E' \subset E_1'$ is evident, we shall prove that if $x \notin E'$, then $x \notin E_1'$ for some E_1. Assume $x \notin E'$, then $x(\xi)$ is not bounded on the unit ball of E. Take $\xi_1 \in E$ such that $x(\xi_1) \geq 1$ and $\|\xi_1\|=1$. Applying mathematical induction, we can choose $\{\xi_1, \xi_2, \cdots, \xi_n, \cdots\}$ which is orthonormal and $x(\xi_k) \geq k^2$. Let E_1 be the subspace spanned by $\{\xi_n\}$, and put $(\xi_j, \xi_k)_1 = k^2 \delta_{jk}$. Then $(\ , \ )$ is evidently HS with respect to $(\ , \ )_1$ on E_1. Furthermore we have $x(\frac{\xi_k}{k}) \geq k$, so that x is not bounded on the unit ball of E_1, hence we get $x \notin E_1'$.

§9. E'-quasi-invariance and E'-ergodicity

Let μ be a measure on $(E^a, \mathcal{B}_E)$, and τ_y be the translation by an element $y \in E^a$: $E^a \ni x \to x+y \in E^a$. Put

$$(9.1) \qquad Y_\mu = \{y \in E^a;\ \tau_y \mu \sim \mu\}.$$

Y_μ is an additive subgroup of E^a, but not a vector space in general.

Theorem 9.1. Let μ be the gaussian measure whose characteristic function is given by (8.2). Then we have $Y_\mu = E'$, where E' is the topological dual space of E in the inner product $(\ , \ )$. In other words, for a subspace $X \subset E^a$, μ is X-quasi-invariant if and only if $X \subset E'$.

Proof. First, we shall show $Y_\mu \subset M^\perp$, where $M=\{\xi \in E;\ (\xi,\xi)=0\}$. If $\xi \in M$, then the distribution of $x(\xi)$ becomes Dirac measure, so that we have $\mu(\{\xi\}^\perp)=1$. Therefore, for every $y \in Y_\mu$ we have

$\{\xi\}^{\perp}\cap(\{\xi\}^{\perp}+y)\neq\phi$, hence $y\in\{\xi\}^{\perp}$. This holds for every $\xi\in M$, thus we get $y\in\bigcap_{\xi\in M}\{\xi\}^{\perp}=M^{\perp}$.

Hereafter we shall assume $y\in M^{\perp}$. Let R be a finite dimensional subspace of E, and μ_R be the restriction of μ on $(E^a,\mathcal{B}_R)$. According to Th.7.1', we have $\tau_y\mu\lesssim\mu$, if and only if $(\tau_y\mu)_R\lesssim\mu_R$ for every R and $(d(\tau_y\mu)_R/d\mu_R)^{1/2}$ converges in $L^2(\mu)$.

Let $\{e_k\}_{k=1}^{n}$ be a base of R such that $\{e_k\}_{k=1}^{m}$ is orthonormal and $(e_k,e_k)=0$ for $m<k\leqq n$. We shall call such a base an orthonormal base of R. Then, the joint distribution of $x(e_k)$ becomes $g_1^m\times\delta^{n-m}$ where g_1 is the one-dimensional gaussian measure with variance 1 and δ is Dirac measure. (cf. Proof of Th.8.1).

For a given $y\in M^{\perp}$, there exists $y_R\in R$ such that $y(\xi)=(\xi,y_R)$ for $\forall\xi\in R$. If $\|y_R\|=0$, then $(\tau_y\mu)_R=\mu_R$. If $\|y_R\|>0$, then put $e_1=y_R/\|y_R\|$ and choose e_k $(k\geqq 2)$ to get an orthonormal base $\{e_k\}$ of R. Since the translation τ_y induces the translation by $\|y_R\|$ only for the first coordinate, the joint distribution of $x(e_k)$ in $\tau_y\mu$ becomes $\tau_{\|y_R\|}g_1\times g_1^{m-1}\times\delta^{n-m}$, therefore we get $(\tau_y\mu)_R\lesssim\mu_R$ and

$$(9.2)\qquad \frac{d(\tau_y\mu)_R}{d\mu_R}(x)=\exp[\tfrac{1}{2}x(e_1)^2-\tfrac{1}{2}(x(e_1)+\|y_R\|)^2]$$

$$=\exp[-x(y_R)-\tfrac{1}{2}\|y_R\|^2].$$

$(d(\tau_y\mu)_R/d\mu_R)^{1/2}$ converges in $L^2(\mu)$ if and only if

$$(9.3)\qquad \left\langle\sqrt{\frac{d(\tau_y\mu)_R}{d\mu_R}},\ \sqrt{\frac{d(\tau_y\mu)_{R_1}}{d\mu_{R_1}}}\right\rangle\to 1.$$

The left hand side of (9.3) is written as

$$\int \exp[-\tfrac{1}{2}x(y_R) - \tfrac{1}{2}x(y_{R_1}) - \tfrac{1}{4}\|y_R\|^2 - \tfrac{1}{4}\|y_{R_1}\|^2]d\mu(x)$$

$$= \exp[-\tfrac{1}{4}(\|y_R\|^2+\|y_{R_1}\|^2)]\int \exp[-\tfrac{1}{2}x(y_R+y_{R_1})]d\mu(x)$$

$$= \exp[-\tfrac{1}{4}(\|y_R\|^2+\|y_{R_1}\|^2)+\tfrac{1}{8}\|y_R+y_{R_1}\|^2]$$

$$= \exp[-\tfrac{1}{8}(\|y_R\|^2+\|y_{R_1}\|^2 - 2(y_R,y_{R_1}))].$$

If $R\subset R_1$, then $(y_R,y_{R_1})=y(y_R)=(y_R,y_R)=\|y_R\|^2$, thus (9.3) is equivalent with

$$\text{(9.4)} \qquad \exp[-\tfrac{1}{8}(\|y_{R_1}\|^2 - \|y_R\|^2)] \to 1,$$

hence with

$$\text{(9.4)'} \qquad \|y_{R_1}\|^2 - \|y_R\|^2 \to 0,$$

hence with the convergence of $\{\|y_R\|\}$.

Since $\|y_R\|=\underset{\xi}{\text{Max}}\{|y(\xi)|;\ \|\xi\|=1,\ \xi\in R\}$, we have $\underset{R}{\text{Sup}}\|y_R\|=\underset{\xi}{\text{Sup}}\{|y(\xi)|;\ \|\xi\|=1\}$, therefore the convergence of $\{\|y_R\|\}$ is equivalent with the boundedness of y on the unit sphere of E, hence with $y\in E'$. Thus (9.3) is satisfied if and only if $y\in E'$. This means $Y_\mu = E'$. (q.e.d.)

For every $\xi\in E$, putting $u_\xi(\eta)=(\xi,\eta)$, we have $u_\xi\in E'$. Then $u_\xi = u_\eta$ is equivalent with $\|\xi-\eta\|=0$, hence with $\xi-\eta\in M$. Thus E/M is imbedded into E'. This imbedding is isometric with respect to the dual norm on E' and the induced norm on E/M from E. Finally E/M is dense in E', because for every $u\in E'$, u_R converges to u in E'. (R is a finite dimensional subspace of E, $u_R\in R$ and $u(\xi)=(\xi,u_R)$ for $\forall\xi\in R$). Thus we get

$$\text{(9.5)} \qquad E' \simeq \overline{E/M}$$

where $\overline{}$ means the completion. Therefore, the inner product $(\ ,\ )$ can be extended on E'. Especially we get $(u,\xi)=u(\xi)$ for $u\in E'$ and $\xi\in E$ (strictly speaking, for $\dot{\xi}\in E/M$).

We can consider another isomorphic representation of E'. Let μ be the gaussian measure corresponding to the inner product $(\ ,\ )$. Then, putting $\Phi_\xi(x)=x(\xi)$, we have $(\xi,\eta)=\langle\Phi_\xi,\Phi_\eta\rangle_{L^2(\mu)}$, so that the map $\xi\to\Phi_\xi$ is isometric. Taking the completion, we get an isomorphism between E' and a closed subspace of $L^2(\mu)$.

$$(9.6) \qquad E' \simeq {}^{\exists}\mathcal{H}_1 \subset L^2(\mu).$$

Thus, every $u\in E'$ corresponds to some $\Phi_u\in L^2(\mu)$. Putting $\Phi_u(x)=(u,x)$, we can consider (u,x) for $u\in E'$ and $x\in E^a$. But note that for a fixed $u\in E'$, (u,x) has a meaning only for almost all $x\in E^a$. (If $\xi\in E$, then $(\xi,x)=x(\xi)$ has a meaning for all $x\in E^a$).

Since the map $u\to\Phi_u$ is linear, we have

$$(9.7) \qquad \forall' x,\ (u+v,x) = (u,x)+(v,x),$$

$$\forall\alpha\in\mathbb{R}^1,\ \forall' x,\ (\alpha u,x) = \alpha(u,x).$$

Here $\forall' x$ means "almost all x". Furthermore we have

$$(9.8) \qquad \forall' x,\ (u,x+v) = (u,x)+(u,v).$$

(If $u(=\xi)\in E$, then (9.8) is evidently valid. For $u\in E'$, we can choose a sequence $\{\xi_n\}\subset E$ such that (ξ_n,x) converges to (u,x) for almost all x. This assures (9.8) for $u\in E'$).

Using the extended inner product, we shall present two

statements. First, for $u \in E'$ put

$$(9.9) \qquad \chi(u) = \int_{E^a} \exp(i(u,x))d\mu(x),$$

then we get

$$(9.10) \qquad \chi(u) = \exp(-\tfrac{1}{2}\|u\|^2).$$

Because, (9.10) is valid for $u(=\xi)\in E$, and both (9.9) and (9.10) are continuous in $\|\cdot\|$.

The second statement is on the density function $d(\tau_u\mu)/d\mu$. From (9.2) and Th.7.3', we have

$$(9.11) \qquad \frac{d(\tau_u\mu)}{d\mu}(x) = \exp(-(u,x)-\tfrac{1}{2}\|u\|^2).$$

Theorem 9.2. Let μ be the gaussian measure corresponding to the inner product (,). For a subspace $X\subset E'$, μ is X-ergodic if and only if X is dense in E'.

Proof. First assume that X is not dense in E'. Then we can find $u(\neq 0)\in E'$ such that $(u,v)=0$ for $\forall v\in X$. From (9.8), the function (u,x) is almost X-invariant, but the distribution of (u,x) becomes the gaussian measure with variance $\|u\|^2$, so that (u,x) is not constant almost everywhere. This means that μ is not X-ergodic.

Next, assume that X is dense in E'. We shall prove that if $f(x)$ is almost X-invariant, then $f(x)$ is almost constant. Without loss of generality, we can assume that $f(x)$ is positive and bounded. Put $d\mu_1=fd\mu$, and denote the characteristic function of μ_1 with χ_1. If $\chi_1(u)=c\chi(u)$ for $\forall u\in E'$, then we get $\mu_1=c\mu$ so that $f(x)=c$ for almost all x. Since χ_1 and χ are continuous in $\|\cdot\|$, it is sufficient to prove $\chi_1(u)=c\chi(u)$

for $u\in X$.

Assume that $u\in X$, $\|u\|=1$. Since $f(x)$ is almost X-invariant, we have $d(\tau_{\lambda u}\mu_1)/d\mu_1=d(\tau_{\lambda u}\mu)/d\mu$ for every $\lambda\in\mathbb{R}^1$. Since $d(\tau_{\lambda u}\mu)/d\mu$ is given by (9.11) replacing u with λu, it is a function of (u,x). Let the distribution of (u,x) in the measure μ_1 be ν, then the distribution of (u,x) in $\tau_{\lambda u}\mu_1$ should be equal with $\tau_\lambda\nu$, so that we have $\frac{d(\tau_\lambda\nu)}{d\nu}(t)=\exp(-\lambda t-\frac{1}{2}\lambda^2)=\frac{d(\tau_\lambda g_1)}{dg_1}(t)$. Thus we see that $d\nu/dg_1$ is $\mathbb{R}^1$-invariant. Since the one-dimensional gaussian measure g_1 is $\mathbb{R}^1$-ergodic, this means that $d\nu/dg_1$ is constant, namely $\nu=cg_1$ for some constant $c>0$. (Here, c is the total measure of μ_1, so it is independent of u). Thus we get

$$\chi(\lambda u) = \int_{-\infty}^{\infty}\exp(i\lambda t)d\nu(t) = c\int_{-\infty}^{\infty}\exp(i\lambda t)dg_1(t) = c\exp(-\tfrac{1}{2}\lambda^2).$$

(q.e.d.)

Remark. The crucial point of the proof is that $d(\tau_{\lambda u}\mu)/d\mu$ is a function of (u,x).

Example. Let μ be the infinite direct product of one-dimensional gaussian measures with variance 1. Then μ is (ℓ^2)-quasi-invariant and $\mathbb{R}_0^\infty$-ergodic.

Let μ be a measure on $(E^a,\mathcal{B}_E)$, and X be a subspace of E^a. If μ is X-quasi-invariant (resp. X-ergodic), then for every $y\in E^a$, the translated measure $\tau_y\mu$ is also X-quasi-invariant (resp. X-ergodic). (cf. Remark 2 of §6).

Especially for the gaussian measure, we get

Theorem 9.3. Let μ be the gaussian measure corresponding to the inner product (,) on E. For every $y\in E^a$, $\tau_y\mu$ is E'-ergodic. We have $\tau_y\mu\sim\tau_z\mu$ if and only if $y-z\in E'$, otherwise we have $\tau_y\mu\perp\tau_z\mu$. (cf. Remark 3 of §6).

In other words, considering the translations of the gaussian measure, we get infinitely many E'-ergodic measures which are mutually singular. They are parametrized by elements of the factor space E^a/E'.

§10. Mutual equivalence

Let (,) and $(,)_1$ be two inner products on E, and μ and μ_1 be the gaussian measures corresponding to them. We shall find the condition for $\mu \sim \mu_1$.

Since $\mu \sim \mu_1$ implies $Y_\mu = Y_{\mu_1}$, the topological dual of E in the norm $\|\cdot\|$ should be equal with that in $\|\cdot\|_1$, therefore two inner products (,) and $(,)_1$ should induce the same topology on E. This means that $\exists a>0, \exists b>0, a\|\xi\| \leq \|\xi\|_1 \leq b\|\xi\|$ for $\forall \xi \in E$. Since $|(\xi,\eta)_1| \leq \|\xi\|_1 \|\eta\|_1 \leq b^2 \|\xi\| \|\eta\|$, there exists a unique linear continuous map A from E to E' which satisfies $(\xi,\eta)_1 = (A\xi,\eta)$. A can be extended to a linear operator from E' to E'. Then A is a positive definite homeomorphic operator.

Theorem 10.1. Let (,) and $(,)_1$ be two inner products on E, and μ and μ_1 be the gaussian measures corresponding to them. We have $\mu \sim \mu_1$, if and only if we can write as $\forall \xi,\eta \in E$, $(\xi,\eta)_1 = (A\xi,\eta)$, where

(1) A is a positive definite homeomorphic linear operator from E' to E', and

(2) I-A is of Hilbert-Schmidt type.

Proof. We have already shown that the condition (1) is necessary. Hereafter we shall always assume (1).

For a finite dimensional subspace R of E, consider the density function $d\mu_{1R}/d\mu_R$ on $\mathcal{B}_R$. If $\mu \sim \mu_1$, then $\sqrt{d\mu_{1R}/d\mu_R}$

should converge to $\sqrt{d\mu_1/d\mu}$ in $L^2(\mu)$, so that

$$(10.1) \qquad \lim_R \int_{E^a} \sqrt{\frac{d\mu_{1R}}{d\mu_R}(x)}\, d\mu(x) > 0.$$

Let $\{e_k\}_{k=1}^n$ be the base of R which is orthonormal in $(\ ,\ )$ and $(Ae_k, e_j) = \lambda_k \delta_{kj}$, then we have

$$(10.2) \qquad \frac{d\mu_{1R}}{d\mu_R}(x) = \prod_{k=1}^n \frac{1}{\sqrt{\lambda_k}} \exp[-\frac{1}{2}(\frac{1}{\lambda_k} - 1)x(e_k)^2].$$

So that the integral in (10.1) is written as

$$\prod_{k=1}^n \lambda_k^{-1/4} \int_{-\infty}^{\infty} \exp[-\frac{1}{4}(\frac{1}{\lambda_k} - 1)t^2 - \frac{t^2}{2}]dt$$
$$= \prod_{k=1}^n \lambda_k^{-1/4} [\frac{1}{2}(\frac{1}{\lambda_k} + 1)]^{-1/2} = [\prod_{k=1}^n \frac{1}{4} \frac{(\lambda_k+1)^2}{\lambda_k}]^{-1/4}$$
$$= [\prod_{k=1}^n (1 + \frac{1}{4} \frac{(\lambda_k-1)^2}{\lambda_k})]^{-1/4}.$$

Thus (10.1) implies $\sup_R \sum_{k=1}^n \frac{(\lambda_k-1)^2}{\lambda_k} < \infty$, hence $\sup_R \sum_{k=1}^n (\lambda_k-1)^2 < \infty$ because $\lambda_k \leqq b^2$ for $\forall k$. This means that $I-A$ is of Hilbert-Schmidt.

Conversely, assume that $I-A$ is of Hilbert-Schmidt, then its eigen vectors form a complete orthonormal system $\{e_k\} \cup \{f_\alpha\}$ in E'.

$$(10.3) \qquad (I-A)e_k = \varepsilon_k e_k, \quad \varepsilon_k \neq 0, \quad \sum_{k=1}^\infty \varepsilon_k^2 < \infty, \quad (I-A)f_\alpha = 0.$$

Let R_n be the space spanned by $\{e_k\}_{k=1}^n$, and μ_{1n} (resp. μ_n) be the restriction of μ_1 (resp. μ) on the smallest σ-algebra in which (e_k, x) is measurable for $1 \leqq k \leqq n$. Then $d\mu_{1n}/d\mu_n$ is given by the right hand side of (10.2), replacing λ_k by $1-\varepsilon_k$ and $x(e_k)$ by (e_k, x). Therefore we get $\langle \sqrt{\frac{d\mu_{1n}}{d\mu_n}}, \sqrt{\frac{d\mu_{1m}}{d\mu_m}} \rangle = [\prod_{k=n+1}^m (1 + \frac{1}{4} \frac{\varepsilon_k^2}{1-\varepsilon_k})]^{-1/4}$, so that $\sum_{k=1}^\infty \varepsilon_k^2 < \infty$

implies that $\sqrt{d\mu_{1n}/d\mu_n}$ converges in $L^2(\mu)$. We shall denote the limit function with $\varphi(x)$. Then if u is a finite linear combination of $\{e_k\}$ and $\{f_\alpha\}$, then we have

$$(10.4)\qquad \int \exp(i(u,x))\varphi^2(x)d\mu(x)$$

$$= \lim_{n\to\infty}\int \exp(i(u,x))d\mu_{1n}(x) = \exp(-\tfrac{1}{2}(Au,u)).$$

From the continuity in $\|\cdot\|$, the equality (10.4) holds even for $\forall u\in E'$ (though the middle term loses its meaning for a general $u\in E'$). This means $d\mu_1=\varphi^2 d\mu$ from the one-to-one correspondence between χ and μ. Hence we have $\mu_1\lesssim\mu$. Since μ and μ_1 are E'-ergodic, we must have $\mu\sim\mu_1$. (q.e.d.)

Let A be a linear map from E into E, and A^* be its adjoint operator from E^a to E^a. For a measure μ on $(E^a,\mathcal{B}_E)$, the characteristic function of the transformed measure $A^*\circ\mu$ is given by $\chi(A\xi)$, because $\int \exp(ix(\xi))d(A^*\circ\mu)(x)= \int \exp(ix(A\xi))d\mu(x)$.

Theorem 10.2. Let (,) be an inner product on E, and μ be the corresponding gaussian measure. If a linear map A from E into E is homeomorphic in the norm $\|\cdot\|$, then A^* maps E' onto E' and A^*A becomes a positive definite homeomorphic operator from E' onto E'. Then the transformed measure $A^*\circ\mu$ is E'-ergodic. We have $A^*\circ\mu\sim\mu$ if and only if $I-A^*A$ is of Hilbert-Schmidt on E'.

Proof is almost evident from Th. 10.1. Note that $A^*\circ\mu$ is the gaussian measure corresponding to $(A\xi,A\eta)=(A^*A\xi,\eta)$.

Let G be the group of all homeomorphic linear maps from E onto E, and G_0 be the subgroup of G which consists of

A such that $I-A^*A$ is of Hilbert-Schmidt. Then, $A^*\circ\mu \sim B^*\circ\mu$ is equivalent with $AB^{-1}\in G_0$. Thus starting from a fixed gaussian measure μ, we get infinitely many E'-ergodic measures $A^*\circ\mu$ which are mutually singular. They are parametrized by elements of $G_0\setminus G$.

Combining with the translations, we get the following result. For $x\in E^a$ and $A\in G$, $\tau_x A^*\circ\mu$ is E'-ergodic. For another pair (y,B), we have $\tau_x A^*\circ\mu \sim \tau_y B^*\circ\mu$ if and only if $x-y\in E'$ and $AB^{-1}\in G_0$, otherwise we have $\tau_x A^*\circ\mu \perp \tau_y B^*\circ\mu$.

§11. Rotationally invariant measures

Let $(\ ,\ )$ be an inner product on E, and μ be the corresponding gaussian measure on $(E^a, \mathcal{B}_E)$. For a homeomorphic linear operator A from E onto E, we have $A^*\circ\mu \sim \mu$ if and only if $I-A^*A$ is of Hilbert-Schmidt (Th.10.2). Meanwhile, we have $A^*\circ\mu = \mu$ if and only if A is isometric (i.e. $I=A^*A$), because the characteristic function of $A^*\circ\mu$ is $\chi(A\xi)$, hence $A^*\circ\mu=\mu$ is equivalent with $\|A\xi\|=\|\xi\|$ for $\forall\xi\in E$.

An isometric linear map from E onto E is called a rotation of E. The group of all rotations of E is called the rotation group of E. Then, the gaussian measure μ is rotationally invariant. $A^*\circ\mu$ is rotationally invariant if and only if A^*A commutes with any rotations, hence $A^*A=cI$ for some constant $c\geq 0$.

Remark 1. Let G be the rotation group of E, and G_0 be a subgroup of G which acts transitively on the unit sphere of E. Then, any G_0-invariant probability measure is G-invariant, because G_0-invariance implies that the characteristic function depends only on $\|\xi\|$.

If $A=cI$, then we have $A^*\circ\mu=\mu_c$ where

$$(11.1) \qquad \forall B\in\mathcal{B}_E,\ \mu_c(B) = \mu(\tfrac{B}{c}).$$

(For $c=0$, μ_0 is the Dirac measure δ).

Theorem 11.1. Let E be infinite dimensional. Every rotationally invariant probability measure m on $(E^a,\mathcal{B}_E)$ is written in the form of

$$(11.2) \qquad \forall B\in\mathcal{B}_E,\ m(B) = \int_{[0,\infty)} \mu_c(B)d\nu(c),$$

where ν is a probability Borel measure on $[0,\infty)$.

Remark 2. Consider the product measure $\mu\times\nu$ on $E^a\times[0,\infty)$, then the right hand side of (11.2) is equal with $(\mu\times\nu)(\Phi^{-1}(B))$ where Φ is a measurable map defined by $\Phi(x,c)=cx$.

Proof (of Th.). Let χ be the characteristic function of m. From the one-to-one correspondence between characteristic functions and measures, it is sufficient to prove that

$$(11.3) \qquad \forall\xi\in E,\ \chi(\xi) = \int_{[0,\infty)} \exp(-\tfrac{1}{2}c^2\|\xi\|^2)d\nu(c).$$

Let ω_n be the uniform probability measure on the unit sphere of $\mathbb{R}^n$. Then the first coordinate x_1 follows the distribution $(1-x_1^2)^{\frac{n-3}{2}}dx_1$ for $|x_1|\leq 1$ (except the normalization constant). Therefore the characteristic function of ω_n is given by

$$(11.4) \qquad \varphi_n(\|\xi\|) = \int_{-1}^{1}\exp(i\|\xi\|t)(1-t^2)^{\frac{n-3}{2}}dt\Big/\int_{-1}^{1}(1-t^2)^{\frac{n-3}{2}}dt.$$

Let $\xi_1,\xi_2,\cdots,\xi_n$ be an orthonormal system of E. Then, the distribution of $(x(\xi_1),\cdots,x(\xi_n))$ in m is rotationally invariant on $\mathbb{R}^n$, so that it is a superposition of uniform measures on the spheres of radius $r \geqq 0$. Thus, the characteristic function χ of m is written as

$$(11.5) \qquad \chi(\xi) = \chi(\|\xi\|\xi_1) = \int_{[0,\infty)} \varphi_n(r\|\xi\|)d\nu_n(r)$$

for some probability measure ν_n on $[0,\infty)$.

Put $\psi_n(\|\xi\|)=\varphi_n(\sqrt{n}\|\xi\|)$, then we have

$$(11.5)' \qquad \chi(\xi) = \int_{[0,\infty)} \psi_n(c\|\xi\|)d\tilde{\nu}_n(c)$$

where $\tilde{\nu}_n$ means the distribution of $\frac{r}{\sqrt{n}}$ in ν_n. On the other hand, we have

$$(11.6) \qquad \psi_n(\|\xi\|) = \int_{-\sqrt{n}}^{\sqrt{n}} \exp(i\|\xi\|t)(1-\frac{t^2}{n})^{\frac{n-3}{2}}dt \Big/ \int_{-\sqrt{n}}^{\sqrt{n}} (1-\frac{t^2}{n})^{\frac{n-3}{2}}dt.$$

Since $\lim_{n\to\infty}(1-\frac{t^2}{n})^{\frac{n-3}{2}}=\exp(-\frac{t^2}{2})$ and since $(1-\frac{t^2}{n})^{\frac{n-3}{2}} \leqq \exp(-\frac{t^2}{8})$ for $n \geqq 4$, from Lebesgue's convergence theorem we see that

$$\lim_{n\to\infty}\int_{-\infty}^{\infty} |(1-\frac{t^2}{n})_+^{\frac{n-3}{2}} - \exp(-\frac{t^2}{2})|dt = 0.$$

(Here, $u_+=u$ for $u>0$, $=0$ for $u \leqq 0$).

Since $|\exp(i\|\xi\|t)| \leqq 1$, $\psi_n(\|\xi\|)$ converges to $\psi(\|\xi\|)$ uniformly in $\|\xi\|$, where

$$\psi(\|\xi\|) = \int_{-\infty}^{\infty} \exp(i\|\xi\|t)\exp(-\frac{t^2}{2})dt \Big/ \int_{-\infty}^{\infty} \exp(-\frac{t^2}{2})dt$$
$$= \exp(-\frac{1}{2}\|\xi\|^2).$$

Therefore from (11.5)', we have

$$(11.7) \qquad \chi(\xi) = \lim_{n\to\infty} \int_{[0,\infty)} \exp(-\frac{c^2}{2}\|\xi\|^2)d\tilde{\nu}_n(c).$$

The set of all probability measures on $[0,\infty]$ is weakly compact, so that some subsequence of $\{\tilde{\nu}_n\}$ has a weak limit ν. For this ν, we get (11.3). (ν is a measure on $[0,\infty]$, but actually it lies on $[0,\infty)$, because χ is continuous in $\|\xi\|$).

(q.e.d.)

Remark 3. Let $\{\xi_k\}_{k=1}^{\infty}$ be an orthonormal system of E. By the strong law of large numbers, we have $\lim_{n\to\infty} \frac{1}{n}\sum_{k=1}^{n} x(\xi_k)^2=1$ for μ-almost all x. Therefore,

$$(11.8) \qquad c(x) = \lim_{n\to\infty}\sqrt{\frac{1}{n}\sum_{k=1}^{n} x(\xi_k)^2}$$

exists for m-almost all x, and ν is the distribution of $c(x)$ in m. (Even if we take another orthonormal system $\{\eta_k\}$ instead of the above $\{\xi_k\}$, $c(x)$ does not change for m-almost all x).

Theorem 11.2. The gaussian measure μ on $(E^a, \mathcal{B}_E)$ is rotationally ergodic. Conversely, a rotationally invariant and rotationally ergodic probability measure should be equal with μ_c for some $c \geqq 0$.

Proof. Let m be rotationally invariant and $m \leqq \mu$. Since $c(x) = 1$ for μ-almost all x, we must have $c(x)=1$ for m-almost all x, hence $\nu(\{1\}^c)=0$ in (11.2). This means $m=a\mu$ for $a=\nu(\{1\})$ (= total measure of ν). Thus, μ is rotationally ergodic. (cf. (E7) in Th.6.1'). In the same way, μ_c is rotationally ergodic for any $c \geqq 0$.

Next, suppose that m is rotationally invariant and rotationally ergodic. Under any rotation, $c(x)$ in (11.8) is invariant for m-almost all x, so that $c(x)$ should be constant m-almost everywhere. (cf. (E5) in Th.6.1). This means that ν is an atomic measure, hence $m=\mu_c$ for some $c \geqq 0$. (q.e.d.)

§12. Representation of $L^2(\mu)$

Let μ be the gaussian measure on $(E^a, \mathcal{B}_E)$ corresponding to an inner product (,) on E. Let $L^2(\mu)$ be the space of all square-integrable functions with respect to μ. Since μ is rotationally invariant, for any rotation U of E, the map $T_U: \Phi(x) \to \Phi(U^*x)$ is a unitary operator on $L^2(\mu)$. Thus $(T_U, L^2(\mu))$ gives a unitary representation of the rotation group of E.

Put

$$(12.1) \qquad \Phi_{\xi_1,\xi_2,\cdots,\xi_n}(x) = \prod_{j=1}^{n} x(\xi_j),$$

and let $\mathcal{M}_n$ be the closed subspace of $L^2(\mu)$ generated by $\{\Phi_{\xi_1,\xi_2,\cdots,\xi_n};\ \xi_1,\xi_2,\cdots,\xi_n \in E\}$. Put $\mathcal{P}_n = \sum_{k=0}^{n} \mathcal{M}_k$ and $\mathcal{H}_n = \mathcal{P}_n \cap \mathcal{P}_{n-1}^{\perp}$. Since each $\mathcal{M}_k$ is rotationally invariant (i.e. $T_U \mathcal{M}_k = \mathcal{M}_k$), we see that $\mathcal{P}_n$, hence $\mathcal{H}_n$ also, is rotationally invariant. Thus $(T_U, \mathcal{H}_n)$ is a unitary representation of the rotation group of E.

$\mathcal{H}_0 = \mathcal{M}_0$ is a one-dimensional space consisting of all constant functions. $\mathcal{H}_1 = \mathcal{M}_1$ is isomorphic with E' (the isomorphism is given by $\Phi_\xi \in \mathcal{H}_1 \leftrightarrow \xi \in E$), as explained in and before (9.6). Since the set of all polynomials is dense in $L^2(\mu)$, $\sum_{n=0}^{\infty} \mathcal{M}_n = \sum_{n=0}^{\infty} \mathcal{H}_n$ is dense in $L^2(\mu)$. Therefore

$$(12.2) \qquad L^2(\mu) = \sum_{n=0}^{\infty} \mathcal{H}_n$$

gives an orthogonal decomposition of $L^2(\mu)$.

Let g be the one-dimensional gaussian measure with variance 1. The functions $1, t, t^2, \cdots, t^n, \cdots$ are linearly independent, so we can consider their Schmidt-orthonormalization $\{h_n(t)\}$

in $L^2(g)$. Namely $h_n(t)$ is a polynomial with degree n and $\langle h_n,h_m\rangle_{L^2(g)}=\delta_{nm}$.

Let $\{u_\alpha\}$ be a complete orthonormal system (= CONS) of E', and put

$$(12.3) \qquad \Phi_{(n_\alpha)}(x) = \prod_\alpha h_{n_\alpha}((u_\alpha,x))$$

for $n_\alpha=0,1,2,\cdots$, $\sum_\alpha n_\alpha<\infty$. ($h_0(t)\equiv 1$ implies that the right hand side of (12.3) is actually a finite product). Then $\{\Phi_{(n_\alpha)}\}$ becomes a CONS of $L^2(\mu)$ and $\{\Phi_{(n_\alpha)};\sum_\alpha n_\alpha=n\}$ becomes a CONS of $\mathcal{H}_n$.

<u>Theorem 12.1</u>. $\{h_n((u,x));\|u\|=1\}$ generates a dense subspace of $\mathcal{H}_n$, and $\langle h_n((u,x)),h_n((v,x))\rangle_{L^2(\mu)}=(u,v)^n$.

<u>Proof</u>. <u>First half</u>. Assume that $\Phi\in\mathcal{H}_n$ and $\langle\Phi,h_n((u,x))\rangle_{L^2(\mu)}=0$ for every $\|u\|=1$. Then $\langle\Phi,(u,x)^n\rangle_{L^2(\mu)}=0$ because $\Phi\perp\mathcal{P}_{n-1}$. Putting $u=\sum_\alpha a_\alpha u_\alpha$ (finite sum, a_α is real), we have $(u,x)^n=\sum_{\Sigma n_\alpha=n}\frac{n!}{\prod n_\alpha!}\prod_\alpha a_\alpha^{n_\alpha}\prod_\alpha(u_\alpha,x)^{n_\alpha}$, hence $\sum_{\Sigma n_\alpha=n}\frac{n!}{\prod n_\alpha!}\prod_\alpha a_\alpha^{n_\alpha}\langle\Phi,\prod_\alpha(u_\alpha,x)^{n_\alpha}\rangle=0$. This holds for every $(a_\alpha)\in\mathbb{R}_0^A$ (A means the set of indices α), so that we have $\langle\Phi,\prod_\alpha(u_\alpha,x)^{n_\alpha}\rangle=0$ for every (n_α) such that $\sum_\alpha n_\alpha=n$. Using again $\Phi\perp\mathcal{P}_{n-1}$, we see that $\langle\Phi,\Phi_{(n_\alpha)}\rangle=0$ for every (n_α) such that $\sum_\alpha n_\alpha=n$, thus we get $\Phi=0$.

<u>Second half</u>. Put $v=u\cos\theta+w\sin\theta$, $u\perp w$, and write $h_n((v,x))$ as a polynomial of (u,x) and (w,x), then the term of the highest degree is $(v,x)^n=\cos^n\theta(u,x)^n+\cdots$, so that we have $\langle h_n((u,x)),h_n((v,x))\rangle=\cos^n\theta=(u,v)^n$ because $h_n((u,x))\perp\mathcal{P}_{n-1}$ and $u\perp w$. (q.e.d.)

<u>Theorem 12.2</u>. The representation $(T_U,\mathcal{H}_n)$ is isomorphic with the symmetric tensor representation $(\otimes^n U,\ S(\otimes^n E'))$.

First we shall explain the latter.

Let X be a vector space and $\otimes^n X$ $(= \underbrace{X\otimes X\otimes\cdots\otimes X}_{n \text{ factors}})$ be its n-ple tensor product. If $\{e_\alpha;\ \alpha\in A\}$ is a base of X, then a base of $\otimes^n X$ is given by $\{e_{\alpha_1}\otimes e_{\alpha_2}\otimes\cdots\otimes e_{\alpha_n};(\alpha_1,\alpha_2,\cdots,\alpha_n)\in A^n\}$. For a linear operator U on X, the linear operator $\otimes^n U$ is defined by

$$(12.4)\qquad \otimes^n U:\ x_1\otimes x_2\otimes\cdots\otimes x_n \to Ux_1\otimes Ux_2\otimes\cdots\otimes Ux_n.$$

Let σ be a permutation of $(1,2,\cdots,n)$. σ induces a linear operator on $\otimes^n X$ by $\sigma(x_1\otimes x_2\otimes\cdots\otimes x_n)=x_{\sigma(1)}\otimes x_{\sigma(2)}\otimes\cdots\otimes x_{\sigma(n)}$. An n-ple tensor $z\in\otimes^n X$ is said to be symmetric if $\sigma z=z$ for every permutation σ of $(1,2,\cdots,n)$. Denote the subspace of all symmetric n-ple tensors with $S(\otimes^n X)$. Evidently $S(\otimes^n X)$ is invariant under $\otimes^n U$.

Let X be a Hilbert space with the inner product $(\ ,\ )$. Define an inner product on $\otimes^n X$ by

$$(12.5)\qquad (x_1\otimes x_2\otimes\cdots\otimes x_n,\ y_1\otimes y_2\otimes\cdots\otimes y_n) = \prod_{k=1}^{n}(x_k,y_k).$$

Hereafter, $\otimes^n X$ means the completion in this inner product, so it is a Hilbert space. If U is a unitary operator on X, then $\otimes^n U$ is unitary on $\otimes^n X$. On the other hand, $S(\otimes^n X)$ is a closed subspace of $\otimes^n X$, because σ is unitary on $\otimes^n X$. Therefore $(\otimes^n U,\ S(\otimes^n X))$ becomes a unitary representation of the unitary group of X, which we shall call the symmetric tensor representation. (For a real Hilbert space, unitarity means rotation).

If $\{e_\alpha;\ \alpha\in A\}$ is a CONS of X, then a CONS of $\otimes^n X$ is given by $\{e_{\alpha_1}\otimes e_{\alpha_2}\otimes\cdots\otimes e_{\alpha_n};\ (\alpha_1,\alpha_2,\cdots,\alpha_n)\in A^n\}$. A CONS of

$S(\otimes^n X)$ is given by $\{\mathcal{E}_{(n_\alpha)};\ n_\alpha=0,1,2,\cdots,\ \sum_\alpha n_\alpha=n\}$, where

$$(12.6)\qquad \mathcal{E}_{(n_\alpha)} = c_{(n_\alpha)}\sum e_{\alpha_1}\otimes e_{\alpha_2}\otimes\cdots\otimes e_{\alpha_n}.$$

Here $\sum$ runs over $(\alpha_1,\alpha_2,\cdots,\alpha_n)$ such that $\forall\alpha\in A,\ \#\{k;\alpha_k=\alpha\}=n_\alpha$, and $c_{(n_\alpha)}$ is the normalization constant given by $c_{(n_\alpha)}=(n!)^{-1/2}(\prod_\alpha n_\alpha!)^{1/2}$.

Similarly as the proof of Th.12.1, we can show that $\{u\otimes u\otimes\cdots\otimes u;\ u\in X\}$ generates a dense subspace of $S(\otimes^n X)$.

Proof of Th.12.2. Put $X=E'$. Since $(u\otimes u\otimes\cdots\otimes u,\ v\otimes v\otimes\cdots\otimes v)=\langle h_n((u,x)),\ h_n((v,x))\rangle=(u,v)^n$, the mapping

$$(12.7)\qquad u\otimes u\otimes\cdots\otimes u \to h_n((u,x)),\ u\in E',\ \|u\| = 1$$

induces an isomorphism between $S(\otimes^n E')$ and $\mathcal{H}_n$. Since $Uu\otimes Uu\otimes\cdots\otimes Uu$ is mapped to $h_n((Uu,x))=h_n((u,U^*x))$, the operator $\otimes^n U$ on $S(\otimes^n E')$ is mapped to T_U on $\mathcal{H}_n$. Thus $(T_U,\mathcal{H}_n)$ is isomorphic with $(\otimes^n U,\ S(\otimes^n E'))$ by the isomorphism defined by (12.7). (q.e.d.)

Remark 1. Putting $u=\sum_\alpha a_\alpha u_\alpha$ $(\sum_\alpha a_\alpha^2=1)$ and comparing the coefficient of $\prod_\alpha a_\alpha^{n_\alpha}$, we see that $\mathcal{E}_{(n_\alpha)}$ in (12.6) is mapped to $\Phi_{(n_\alpha)}$ in (12.3) by the isomorphism (12.7). (In (12.6), e_α should be replaced by u_α).

Remark 2. Special case of $E=L^2(\Omega,\nu)$ for some measure space (Ω,ν).

Consider the product measure space (Ω^n,ν^n), then we have $\otimes^n L^2(\Omega,\nu)\simeq L^2(\Omega^n,\nu^n)$. Here the isomorphism is given by $f_1\otimes f_2\otimes\cdots\otimes f_n\leftrightarrow f_1(\omega_1)f_2(\omega_2)\cdots f_n(\omega_n)$. Then $S(\otimes^n L^2(\Omega,\nu))$ corresponds to the set of all symmetric functions in $L^2(\Omega^n,\nu^n)$. Here a symmetric function means $\varphi(\omega_1,\omega_2,\cdots,\omega_n)=\varphi(\omega_{\sigma(1)},\omega_{\sigma(2)},\cdots,\omega_{\sigma(n)})$ for every permutation σ.

Let V be a measure-preserving map on Ω. Then the map $V^n: (\omega_1,\omega_2,\cdots,\omega_n)\to(V\omega_1,V\omega_2,\cdots,V\omega_n)$ is a measure-preserving map on Ω^n. Put $U=T_V: f(\omega)\to f(V\omega)$ on $L^2(\Omega,\nu)$, then $\otimes^n T_V$ on $S(\otimes^n L^2(\Omega,\nu))$ is mapped to T_{V^n} on $L^2(\Omega^n,\nu^n)$. So that, for the case of $E=L^2(\Omega,\nu)$, we have $(T_U,\mathcal{H}_n)\simeq(\otimes^n U,S(\otimes^n L^2(\Omega,\nu)))\simeq (T_{V^n}, S(L^2(\Omega^n,\nu^n)))$ for $U=T_V$.

Theorem 12.3. The symmetric tensor representation $(\otimes^n U, S(\otimes^n X))$ of the unitary group of X is irreducible. $(\otimes^n U, S(\otimes^n X))$ is not equivalent with $(\otimes^m U, S(\otimes^m X))$ for $n\neq m$. Therefore, $(T_U,\mathcal{H}_n)$ is irreducible, and $(T_U,\mathcal{H}_n)$ is not equivalent with $(T_U,\mathcal{H}_m)$ for $n\neq m$.

Proof. First half. Let T be a bounded operator on $S(\otimes^n X)$. $(\otimes^n U, S(\otimes^n X))$ is irreducible, if $T(\otimes^n U)=(\otimes^n U)T$ for every U implies that T should be a constant multiplier, namely that $\exists\lambda$, $\forall x\in X$, $T(\otimes^n x)=\lambda\otimes^n x$, where $\otimes^n x$ means the n-ple tensor product of the same vector x.

Without loss of generality, we can assume $\|x\|=1$. Let $\{u_\alpha;\alpha\in A\}$ be a CONS of X including x as e_0, and put $T(\otimes^n e_0)=\sum_{(n_\alpha)} a_{(n_\alpha)}\mathcal{E}_{(n_\alpha)}$, where $\mathcal{E}_{(n_\alpha)}$ is the CONS of $S(\otimes^n X)$ given by (12.6).

Let σ be a permutation of A, then $U_\sigma: e_\alpha\to e_{\sigma(\alpha)}$ induces a unitary operator U_σ on X. Since T commutes with $\otimes^n U_\sigma$, if $\sigma(0)=0$, then we have $a_{(n_\alpha)}=a_{(n'_\alpha)}$ for $n'_\alpha=n_{\sigma(\alpha)}$. When we change σ over all permutations of $A-\{0\}$, we get infinitely many different (n'_α)s except for $(n_\alpha)=(n,0,0,\cdots)$. Since $\sum_{(n_\alpha)} a^2_{(n_\alpha)}<\infty$, we see that $a_{(n_\alpha)}=0$ except for $(n_\alpha)=(n,0,0,\cdots)$. This means that $T(\otimes^n e_0)=\lambda\otimes^n e_0$. ($\lambda$ is independent of the choice of $e_0=x$, because T commutes with $\otimes^n U$).

Second half. Let T be a bounded linear map from $S(\otimes^n X)$ to $S(\otimes^m X)$. We shall show that $T(\otimes^n U)=(\otimes^m U)T$ for every U implies $T=0$. Similarly with the proof of First half, we have $\exists\lambda$, $\forall x\in X$, $T\otimes^n x=\lambda\otimes^m x$. Let T^* be the adjoint map of T from $S(\otimes^m X)$ to $S(\otimes^n X)$. Evidently we have $T^*(\otimes^m x)=\lambda\otimes^n x$. Then $(T\otimes^n x,\otimes^m y)=(\otimes^n x,T^*(\otimes^m y))$ implies $\lambda(x,y)^m=\lambda(x,y)^n$. This holds for every $|(x,y)|\leqq 1$, so that λ should be zero. (q.e.d.)

Corollary 1. Let A be a rotationally invariant bounded operator on $L^2(\mu)$. (This means $AT_U=T_UA$ for every rotation U of E). Then, A is a constant multiplier on each $\mathcal{H}_n$.

Let P_m be the projection onto $\mathcal{H}_m$. The restriction of P_mA on $\mathcal{H}_n$ is a bounded map from $\mathcal{H}_n$ to $\mathcal{H}_m$, so that from Th.12.3, on the space $\mathcal{H}_n$, $P_mA=0$ for $m\neq n$ and P_nA is a constant multiplier. Therefore, A is a constant multiplier on $\mathcal{H}_n$.

Corollary 2. Let A be a rotationally invariant Hermite operator on $L^2(\mu)$. Then, each $\mathcal{H}_n$ is contained in the domain of A and A is a constant multiplier on each $\mathcal{H}_n$.

Here we shall give an example of a rotationally invariant Hermite operator.

As well known, the Laplacian operator Δ_n is Hermitian on $L^2(\mathbb{R}^n, d^nx)$, where d^nx means the Lebesgue measure on $\mathbb{R}^n$. Let g be the one-dimensional gaussian measure with variance 1, then $L^2(\mathbb{R}^n, d^nx)$ is isomorphic with $L^2(\mathbb{R}^n, g^n)$ by $\varphi(x)\overset{T}{\leftrightarrow}(2\pi)^{n/4}\exp[\frac{1}{4}\sum_{k=1}^{n} x_k^2]\varphi(x)$. Then Δ_n is mapped to $T\Delta_nT^{-1}=\sum_{k=1}^{n}\frac{\partial^2}{\partial x_k^2}-\sum_{k=1}^{n}x_k\frac{\partial}{\partial x_k}-\frac{n}{2}+\frac{1}{4}\sum_{k=1}^{n}x_k^2$. Since $-\frac{n}{2}+\frac{1}{4}\sum_{k=1}^{n}x_k^2$ is Hermitian, we see that $\sum_{k=1}^{n}\frac{\partial^2}{\partial x_k^2}-\sum_{k=1}^{n}x_k\frac{\partial}{\partial x_k}$ is Hermitian on $L^2(\mathbb{R}^n,g^n)$.

Put $\mathcal{P}=\{F(x)\in L^2(\mu);\ ^{\exists}P$ polynomial, $^{\exists}\xi_1,\xi_2,\cdots,\xi_n\in E$, $F(x)=P(x(\xi_1),x(\xi_2),\cdots,x(\xi_n))\}$. Here $\{\xi_k\}$ can be assumed to be orthonormal. Put

(12.8) $$\Delta_\infty F(x) = \sum_{k=1}^{n}\left(\frac{\partial^2}{\partial t_k^2} - t_k\frac{\partial}{\partial t_k}\right)P(t_1,t_2,\cdots,t_n)\Big|_{t_j=x(\xi_j)}$$

where $\Big|_{t_j=x(\xi_j)}$ means the substitution after the calculation.

It is easily checked that Δ_∞ is independent of the representation of $F(x)$ as $P(x(\xi_1),x(\xi_2),\cdots,x(\xi_n))$, $\{\xi_k\}$ being orthonormal. Thus Δ_∞ is defined on $\mathcal{P}$, and is called the infinite dimensional Laplacian operator. Δ_∞ is extended to a Hermite operator on $L^2(\mu)$.

Let $\mathbb{P}_n$ be the set of all polynomials of a single variable with degree $\leq n$. The operator $\frac{d^2}{dt^2} - t\frac{d}{dt}$ keeps $\mathbb{P}_n$ invariant. Since this operator is Hermitian on $L^2(\mathbb{R},g)$, its eigen functions are orthogonal with each other. So that, the function $h_n(t)$ defined before (12.3) is the eigen function. Since $h_n(t)$ is of degree n, the eigen value should be $-n$.

From this, we see that $\mathcal{H}_n$ is the eigen space of Δ_∞ belonging to the eigen value $-n$.

Remark 3. $h_n(t)$ is the eigen function of $\frac{d^2}{dt^2} - t\frac{d}{dt}$. On the other hand, the eigen function of $\frac{d^2}{dt^2} - 2t\frac{d}{dt}$ is known as Hermite polynomial $H_n(t)$. So, using Hermite polynomial, we have

(12.9) $$h_n(t) = c_n H_n\left(\frac{t}{\sqrt{2}}\right)$$

where c_n is the normalization constant given by $c_n=2^{-n/2}(n!)^{-\frac{1}{2}}$. As well known, $H_n(t)$ is also written as

(12.10) $$H_n(t) = (-1)^n \exp t^2 \frac{d^n}{dt^n}\exp(-t^2).$$

CHAPTER 3. THE SET Y_μ OF ALL QUASI-INVARIANT TRANSLATIONS

§13. Convolution of measures

Let X be a vector space, Y be a subspace of X and R be a subspace of X', the algebraical dual space of X. Let $\mathcal{B}_R$ be the smallest σ-algebra on X in which every $f \in R$ is measurable. A measure on $(X, \mathcal{B}_R)$ is said to be Y-quasi-invariant if it is quasi-invariant under translations τ_y: $x \to x+y$ for $y \in Y$. Put

$$(13.1) \qquad Y_\mu = \{y \in X;\ \mu \text{ is quasi-invariant under } x \to x+y\},$$

then μ is Y-quasi-invariant if and only if $Y \subset Y_\mu$.

As proved in Chapter 1, if X is infinite dimensional, there exists no X-quasi-invariant measure on $(X, \mathcal{B}_R)$. Our problem in Chapter 3 is: when does a Y-quasi-invariant measure exist?

<u>Definition 13.1</u>. Let μ and ν be two measures on $(X, \mathcal{B}_R)$. The following measure $\mu * \nu$ is called the convolution of μ and ν.

$$(13.2) \qquad \forall E \in \mathcal{B}_R,\ (\mu * \nu)(E) = \int_X \mu(E-x)\,d\nu(x).$$

In other words, putting $\Phi: (x,y) \to x+y$, we get $(\mu*\nu)(E) = (\mu \times \nu)(\Phi^{-1}(E))$. It is easy to see $\mu*\nu = \nu*\mu$.

If we denote the characteristic functions of μ, ν and $\mu*\nu$ by χ_μ, χ_ν and $\chi_{\mu*\nu}$ respectively, we can easily show

$$(13.3) \qquad \chi_{\mu*\nu}(\xi) = \chi_\mu(\xi)\chi_\nu(\xi), \quad \forall \xi \in R.$$

If $\mu \sim \mu'$ and $\nu \sim \nu'$, then we have $\mu*\nu \sim \mu'*\nu'$.

Theorem 13.1. $Y_{\mu*\nu} \supset Y_\mu + Y_\nu$.

Proof. Put

(13.4) $\mu_y(E) = \mu(E-y), \forall E \in \mathcal{B}_R$.

Then $y \in Y_\mu$ is equivalent with $\mu \sim \mu_y$. From (13.2) we can deduce $(\mu*\nu)_y = \mu_y*\nu = \mu*\nu_y$. Therefore we get both $Y_\mu \subset Y_{\mu*\nu}$ and $Y_\nu \subset Y_{\mu*\nu}$. Since $Y_{\mu*\nu}$ is an additive group, this means that $Y_{\mu*\nu} \supset Y_\mu + Y_\nu$.

(q.e.d)

Theorem 13.2. Suppose that X is infinite dimensional. If Y_1, $Y_2, \cdots, Y_n$ are subspaces of X such that $Y_1+Y_2+\cdots+Y_n=X$, then there exists no Y_i-quasi-invariant measure for some i.

Proof. If μ_i is a Y_i-quasi-invariant measure for each i, then $\mu_1*\mu_2*\cdots*\mu_n$ is X-quasi-invariant. This contradicts to Th.5.3.

(q.e.d)

Corollary. If X is infinite dimensional and Y is finite codimensional (namely if $X=Y+Z$ for some finite dimensional Z), then no Y-quasi-invariant measure exists.

Proof. The Lebesgue measure on Z can be imbedded as a measure on $(X, \mathcal{B}_R)$. Since this measure is Z-invariant, there exists no Y-quasi-invariant measure according to Th.13.2. (q.e.d)

§14. Linearization of a topology on a vector space

This section is a preliminary for the next section. It concerns a general discussion on a vector topology.

Theorem 14.1. Let τ be a topology on a vector space X. Suppose that the addition is continuous in τ (namely (X,τ) is a topological additive group), and suppose that for every fixed $x \in X$, the map $\mathbb{R} \ni \alpha \to \alpha x \in X$ is continuous in τ. This means that denoting the fundamental system of neighbourhoods of 0 by $\{V\}=\mathcal{V}$, we have

(14.1) $\quad \forall x\in X,\ \forall V\in\mathcal{V},\ \exists \alpha_0>0,\ |\alpha|\leq\alpha_0 \Rightarrow \alpha x\in V.$

Put

(14.2) $\quad U_V = \bigcap_{|\alpha|\geq 1} (\alpha V),$

then $\{U_V\}_{V\in\mathcal{V}}$ is a fundamental system of neighbourhoods of 0 in a topology τ'. This topology τ' is the weakest one which is stronger than τ and compatible with the vector operations on X (namely makes X a topological vector space).

Definition 14.1. The topology τ' in Th. 14.1 is called the linearization of τ.

Proof. Consider a vector topology τ'' which is stronger than τ. Then we have $\forall V\in\mathcal{V}$ $\exists\alpha_0>0$, $\exists U$ (neighbourhood of 0 in τ''), $|\alpha|\leq\alpha_0 \Rightarrow \alpha U\subset V$, hence $\alpha_0 U\subset U_V$. Since $\alpha_0 U$ is a neighbourhood of 0 in τ'', this means that τ'' is stronger than τ'.

Now, we shall prove that X is a topological vector space in τ'. Since $V+V\subset W$ implies $U_V+U_V\subset U_W$, the addition is continuous in τ'. (Of course we define a fundamental system of neighbourhoods of x by $\{x+U_V\}_{V\in\mathcal{V}}$). In order to prove that the scalar multiplication is continuous, we need to check the following three points:

1) $\quad \forall V\in\mathcal{V},\ \exists\alpha_0>0,\ \exists V_0\in\mathcal{V},\ |\alpha|\leq\alpha_0 \Rightarrow \alpha U_{V_0}\subset U_V,$

2) $\quad \forall\alpha\neq 0,\ \forall V\in\mathcal{V},\ \exists V_0\in\mathcal{V},\ \alpha U_{V_0}\subset U_V,$

3) $\quad \forall x\neq 0,\ \forall V\in\mathcal{V},\ \exists\alpha_0>0,\ |\alpha|\leq\alpha_0 \Rightarrow \alpha x\in U_V.$

From (14.2), we have $\alpha U_V = \bigcap_{|\beta|\geq 1} (\beta\alpha V) = \bigcap_{|\beta|\geq|\alpha|} (\beta V)$, so that $|\alpha|\leq 1$ implies $\alpha U_V \subset U_V$. This proves 1) by choosing $V_0 = V$ and $\alpha_0 = 1$.

Suppose that $\alpha\neq 0$ is given, and choose an integer n such that $|\alpha|<n$. Since the addition is continuous in τ, we have $\forall V\in\mathcal{V}$, $\exists V_n\in\mathcal{V}$, $\underbrace{V_n+V_n+\cdots+V_n}_{n\text{ terms}}\subset V$, so especially we have $nV_n\subset V$. From this we get $U_V = \bigcap_{|\beta|\geq 1} (\beta V) \supset \bigcap_{|\beta|\geq 1} (n\beta V_n) = \bigcap_{|\beta|\geq n} (\beta V_n) \supset \bigcap_{|\beta|\geq|\alpha|} (\beta V_n) = \alpha U_{V_n}$. This proves 2).

The assumption (14.1) is equivalent with $\forall x\in X$, $\forall V\in\mathcal{V}$, $\exists\alpha_0>0$, $\alpha_0 x\in U_V$. So, if $|\alpha|\leq\alpha_0$ we have $\alpha x = \frac{\alpha}{\alpha_0}(\alpha_0 x) \in \frac{\alpha}{\alpha_0} U_V \subset U_V$. This completes the proof of 3). (q.e.d.)

§15. Characteristic topology

Let X be a vector space, R be a subspace of X' and μ be a probability measure on $(X, \mathcal{B}_R)$. Denote the characteristic function of μ(defined on R) with $\chi(\xi)$.

<u>Theorem 15.1</u>. There exists the weakest vector topology on R in which $\chi(\xi)$ is continuous.

<u>Proof</u>. Put

$$V_\varepsilon = \{\xi\in R;\ |1-\chi(\xi)|\leq\varepsilon\}. \tag{15.1}$$

Since $\chi(\xi)$ satisfies $|\chi(\xi+\eta)-\chi(\xi)|\leq\sqrt{2|1-\chi(\eta)|}$, (c.f. Part A (16.6)), we have

$$V_\varepsilon + V_\varepsilon \subset V_{\varepsilon+\sqrt{2\varepsilon}}. \tag{15.2}$$

Thus X is a topological additive group in the topology τ whose fundamental system of neighbourhoods of 0 is given by $\{V_\varepsilon\}_{\varepsilon>0}$. For every fixed $\xi\neq 0$, $\chi(\alpha\xi)$ is a continuous function of α, because it is the characteristic function of the image measure of μ by $X\ni x\to\xi(x)\in\mathbb{R}^1$. Thus we have

(15.3) $\forall\xi\in R,\ \forall\varepsilon>0,\ \exists\alpha_0>0,\ |\alpha|\leqq\alpha_0\Rightarrow\alpha\xi\in V_\varepsilon.$

By Th. 14.1, we can consider the linearization of τ. The linearized topology τ' is evidently the requestted one in the theorem. (q.e.d.)

Definition 15.1. The topology mentioned in Th. 15.1 is called the characteristic topology of μ and denoted with τ_μ.

The fundamental system of neighbourhoods of 0 in τ_μ is given by $\{U_\varepsilon\}_{\varepsilon>0}$ where

(15.4) $U_\varepsilon = \{\xi\in R;\ \sup_{|\alpha|\leqq 1}|1-\chi(\alpha\xi)|\leqq\varepsilon\}.$

τ_μ is not necessarily Hausdorff. We have $\xi\in\bigcap_{\varepsilon>0}U_\varepsilon$ if and only if $\xi(x)=0$ for μ-almost all x. In other words, τ_μ becomes a Hausdorff topology on the factor space R/M, where $M=\{\xi\in R;\ \xi(x)=0$ for μ-almost all $x\}$.

Theorem 15.2. Suppose that τ is a vector topology on R. $\chi(\xi)$ is continuous in τ if and only if τ is stronger than τ_μ.

Proof is evident from the definition of τ_μ.

Next, we shall prove that the characteristic topology is identical with the topology of measure convergence. As a preliminary discussion, we shall explain the latter.

Let $(X,\mathcal{B},\mu)$ be a probability measure space, and $\mathcal{F}$ be the vector space of all $\mathcal{B}$-measurable functions on X. The sequence $f_n\in\mathcal{F}$ is said to converge to 0 in measure μ if

(15.5) $\forall\alpha>0,\ \mu(\{x\in X;\ |f_n(x)|\geqq\alpha\})\to 0\quad(n\to\infty).$

We can formulate this using a topology on $\mathcal{F}$. Put

(15.6) $$U_{\alpha,\varepsilon} = \{f\in\mathcal{F};\ \mu(\{x\in X;\ |f(x)|\geq\alpha\})<\varepsilon\}\ ,$$

then $\{U_{\alpha,\varepsilon}\}_{\alpha>0,\varepsilon>0}$ becomes a fundamental system of neighbourhoods of 0 in some topology. This topology is called the topology of measure convergence in μ.

$U_{\alpha,\varepsilon}$ is increasing with respect to both α and ε, so that $\{U_{\frac{1}{n},\frac{1}{n}}\}_{n=1,2\cdots}$ is a fundamental system of neighbourhoods of 0. We can easily check $U_{\alpha,\varepsilon}+U_{\beta,\delta}\subset U_{\alpha+\beta,\varepsilon+\delta}$, so that $\mathcal{F}$ is a topological additive group. We can also check $cU_{\alpha,\varepsilon}=U_{|c|\alpha,\varepsilon}$, so combining this with $\forall f\in\mathcal{F}$, $\forall\varepsilon>0$, $\exists\alpha>0$, $f\in U_{\alpha,\varepsilon}$, we see that the scalar multiplication is also continuous. Namely $\mathcal{F}$ becomes a topological vector space in the topology of measure convergence.

$f\in\bigcap_{\alpha>0,\varepsilon>0} U_{\alpha,\varepsilon}$ is equivalent with $f(x)=0$ for μ-almost all x. In other words, the topology of measure convergence is Hausdorff on the factor space $\mathcal{F}/\mathfrak{M}$ where $\mathfrak{M}=\{f\in\mathcal{F};\ f(x)=0$ for μ-almost all $x\}$. The factor space $\mathcal{F}/\mathfrak{M}$ is often denoted with $L^0(\mu)$.

We shall give another formulation of the topology of measure convergence. Let $F(t)$ be a bounded continuous function on $\mathbb{R}^1$. For $f\in\mathcal{F}$, consider

(15.7) $$I_F(f) = \int_X F(f(x))d\mu(x).$$

Lemma 15.1. 1) $I_F(f)$ is continuous at $f=0$ in the topology of measure convergence.

2) If we assume that

(15.8) $$\forall\alpha>0,\ \inf_{|t|\geq\alpha} F(t)>F(0),$$

then $I_F(f_n)\to I_F(0)$ implies that $\{f_n\}$ converges to 0 in measure μ.

Proof. Since $F(t)$ is continuous, for a given $\varepsilon>0$, there exists $\delta>0$ such that $|t|\leq\delta$ implies $|F(t)-F(0)|\leq\varepsilon$. Then, we have $|I_F(f)-I_F(0)|\leq\int_X|F(f(x))-F(0)|d\mu(x)\leq\varepsilon\mu(f^{-1}(-\delta,\delta))+2M\mu(\{x;\ |f(x)|\geq\delta\})$, where $M=\sup_t|F(t)|$. Therefore $f\in U_{\delta,\varepsilon/2M}$ implies $|I_F(f)-I_F(0)|\leq 2\varepsilon$. This proves 1).

Now, assume (15.8). In this case we have $I_F(f)-I_F(0)=\int_X(F(f(x))-F(0))d\mu(x)\geq(\inf_{|t|\geq\alpha}F(t)-F(0))\mu(\{x;\ |f(x)|\geq\alpha\})$. Therefore $I_F(f)-I_F(0)<\varepsilon(\inf_{|t|\geq\alpha}F(t)-F(0))$ implies $f\in U_{\alpha,\varepsilon}$. This proves 2). (q.e.d.)

Thus, if a bounded continuous function $F(t)$ satisfies (15.8), the fundamental system of neighbourhoods of 0 in the topology of measure convergence is given by $\{W_\varepsilon\}_{\varepsilon>0}$, where

$$(15.9)\qquad W_\varepsilon = \{f;\ I_F(f)<I_F(0)+\varepsilon\}.$$

Especially if we put $F(t)=\frac{|t|}{1+|t|}$, the corresponding $I_F(f)$ gives a metric on $\mathfrak{F}$. Namely the topology of measure convergence is defined by the metric:

$$(15.10)\qquad d(f,g) = \int_X \frac{|f(x)-g(x)|}{1+|f(x)-g(x)|}\,d\mu(x).$$

Of course we can consider another metric by choosing suitable $F(t)$.

Theorem 15.3. Let μ,ν be two probability measures on a measurable space $(X,\mathcal{B})$. If $\nu\prec\mu$ (namely ν is absolutely continuous with respect to μ), then the topology of measure convergence in μ is stronger than that in ν.

Corollary. If $\nu\sim\mu$, then two topologies of measure convergence are identical.

Proof of Th. Since the neighbourhoods of 0 are given by (15.6), it is sufficient to prove that

(15.11) $\quad \forall \varepsilon>0,\ \exists \delta>0,\ \mu(B)<\delta \Rightarrow \nu(B)<\varepsilon.$

We shall prove the contraposition. Assume that $\exists \varepsilon>0$, $\forall n$, $\exists B_n \in \mathcal{B}$, $\mu(B_n)<\frac{1}{2^n}$ but $\nu(B_n)\geqq\varepsilon$, then putting $B=\overline{\lim}\, B_n$, we have $\mu(B)=0$ but $\nu(B)\geqq\varepsilon$. This contradicts to $\nu\lesssim\mu$. (q.e.d.)

We shall consider again the case of a vector space.

Let X be a vector space, R be a subspace of X' and μ be a probability measure on $(X,\mathcal{B}_R)$. Since every $\xi\in R$ is $\mathcal{B}_R$-measurable, we can imbed R into $\mathcal{F}$, the space of all $\mathcal{B}_R$-measurable functions on X.

Theorem 15.4. The characteristic topology of μ is identical with the restriction of the topology of measure convergence on R.

Proof. The former is weaker than the latter, because the characteristic function $\chi(\xi)$ is continuous in the topology of measure convergence. This can be shown from 1) of Lemma 15.1, since $\chi(\xi)=I_{F_1}(\xi)+iI_{F_2}(\xi)$ where $F_1(t)=\cos t$ and $F_2(t)=\sin t$.

The former is stronger than the latter, because if $\xi\in U_\varepsilon$ in (15.4), then we get

$$\varepsilon\geqq\frac{1}{2}\int_{-1}^{1}(1-\chi(\alpha\xi))d\alpha=\int_X\left(1-\frac{\sin\xi(x)}{\xi(x)}\right)d\mu(x),$$

thus ξ is near to 0 in the topology of measure convergence as seen from 2) of Lemma 15.1 putting $F(t)=1-\frac{\sin t}{t}$. (q.e.d.)

Corollary of Th. 15.4. Let μ and ν be two probability measures on $(X,\mathcal{B}_R)$. If $\nu\lesssim\mu$, then the characteristic topology of μ is stronger than that of ν. Especially if $\mu\sim\nu$, then two characteristic topologies are identical with each other.

§16. Evaluation of Y_μ in terms of τ_μ

Let X be a vector space and R be a subspace of X',

satisfying the condition $\forall x \neq 0, \exists \xi \in R, \xi(x) \neq 0$. Then X can be imbedded into R', and any measure on $(X, \mathcal{B}_R)$ can be regarded as a measure on R'. Hereafter we shall consider a measure always on $(R', \mathcal{B}_R)$ unless otherwise mentioned. Then, Y_μ is a subset of R' and τ_μ is a topology on R.

Theorem 16.1. We have $Y_\mu \subset R^*_\mu$, where R^*_μ is the topological dual space of R in the topology τ_μ.

Remark. Though τ_μ may not be locally convex, we can define R^*_μ as the set of all linear continuous functions on R in τ_μ.

Proof. Suppose that $y \in Y_\mu (\subset R')$. Since $\mu_y \sim \mu$, the characteristic topology of μ_y is identical with that of μ. This implies that $\exp(iy(\xi))\chi(\xi)$ is continuous in τ_μ, so that $\exp(iy(\xi))$ is continuous in τ_μ at $\xi = 0$. Therefore for some neighbourhood U of 0 in τ_μ, we have $\sup_{\xi \in U} |1 - \exp(iy(\xi))| \leq 1$, hence $\sup_{\xi \in U} |y(\xi)| \leq \pi/3$. This implies that $y(\xi)$ is continuous in τ_μ, hence $y \in R^*_\mu$. (q.e.d.)

Corollary. If $\chi(\xi)$ is continuous in a vector topology τ, then we have $Y_\mu \subset R^*_\tau$, the topological dual space of R in τ.

Proof. Since τ is stronger than τ_μ, we have $R^*_\mu \subset R^*_\tau$. Combining this with Th. 16.1, we get the corollary. (q.e.d.)

We shall define $\|x\|_A$ $(x \in R', A \subset R)$ as follows:

$$\|x\|_A = \sup_{\xi \in A} |x(\xi)|. \tag{16.1}$$

Similarly define $\|\xi\|_{A'}$ $(\xi \in R, A' \subset R')$ as follows:

$$\|\xi\|_{A'} = \sup_{x \in A'} |x(\xi)|. \tag{16.2}$$

$\|\cdot\|_A$ (resp. $\|\cdot\|_{A'}$) is a semi-norm on the space $\{x \in R'; \|x\|_A < \infty\}$ (resp. $\{\xi \in R; \|\xi\|_{A'} < \infty\}$).

Theorem 16.2. Let Y be a subspace of R' such that Y admits a complete metrizable vector topology stronger than the weak topology. If $Y \subset Y_\mu$ (namely if μ is Y-quasi-invariant), then for some neighbourhood V of 0 in Y, the semi-norm $\|\cdot\|_V$ is finite on R and τ_μ is stronger than the semi-normed topology induced from $\|\cdot\|_V$.

Proof. Let $\{U_n\}_{n=1,2,\dots}$ be a fundamental system of neighbourhoods of 0 in τ_μ, and put

$$(16.3) \qquad \begin{aligned} \mathring{U}_n &= \{x \in R';\ \|x\|_{U_n} \leqq 1\} \\ &= \{x \in R';\ |x(\xi)| \leqq 1 \quad \text{for } \forall \xi \in U_n\}. \end{aligned}$$

Evidently $\mathring{U}_n$ is weakly closed and $\bigcup_{n=1}^{\infty} \mathring{U}_n = R^*_\mu$. Combine the assumption $Y \subset Y_\mu$ with the result of Th. 16.1, then we see $Y \subset R^*_\mu$, hence we get $Y = \bigcup_{n=1}^{\infty} (Y \cap \mathring{U}_n)$. Since the topology of Y is stronger than the weak topology, $Y \cap \mathring{U}_n$ is a closed set. Since Y is complete and metrizable, by Baire's theorem some $Y \cap \mathring{U}_n$ must have an inner point. This means $\exists y_0 \in Y$, $\exists V$ (neighbourhood of 0 in Y), $y_0 + V \subset \mathring{U}_n$. Therefore we have $V \subset \mathring{U}_n - \mathring{U}_n \subset 2\mathring{U}_n$. From this we get $\mathring{V} \supset \frac{1}{2} U_n$ (where $\mathring{V} = \{\xi \in R;\ \|\xi\|_V \leqq 1\}$). Thus $\|\xi\|_V$ is finite on R and τ_μ is stronger than the topology induced from $\|\cdot\|_V$. (q.e.d.)

Theorem 16.3. Let τ be a locally convex vector topology on R. Then $Y_\mu \supset R^*_\tau$ implies that τ is weaker than τ_μ.

Proof. Since τ is locally convex, τ is defined by a family of semi-norms $\{p_\lambda\}$. So, it is sufficient to show that each p_λ is continuous in τ_μ.

Let p be a continuous semi-norm in τ, and R^*_p be the topological dual space of R in the semi-norm p. Evidently we have $R^*_p \subset R^*_\tau$ and R^*_p is a Banach space with respect to the dual norm p^*. Therefore $Y_\mu \supset R^*_\tau$ implies $Y_\mu \supset R^*_p$, so that by Th. 16.2 τ_μ is stronger than the dual topology of p^*, which is identical with p. Thus p is continuous in τ_μ. (q.e.d.)

Combining Th. 16.3 with the Corollary of Th. 16.1, we see that for a locally convex topology τ on R

τ is stronger than $\tau_\mu \Rightarrow Y_\mu \subset R^*_\tau$, and

$Y_\mu \supset R^*_\tau \Rightarrow \tau$ is weaker than τ_μ.

From this we get

Theorem 16.4. Let τ be a locally convex topology on R. If $\chi(\xi)$ is continuous in τ and if μ is R^*_τ-quasi-invariant, then we have $\tau=\tau_\mu$ and $Y_\mu=R^*_\tau$.

Proof. The assumptions are equivalent with that τ is stronger than τ_μ and $Y_\mu \supset R^*_\tau$. So the result is evident. (q.e.d.)

Remark. Actually, unless (R,τ) is a Hilbert space, no measure satisfies the assumptions in Th. 16.4. (H. Shimomura: Journal of Mathematics of Kyoto University, Vol.21, No.4 (1981) 703-713).

§17. Some applications

Let μ be a probability measure on $(R',\mathcal{B}_R)$. Let $A_n \in \mathcal{B}_R$ be such that $\lim_{n\to\infty}\mu(A_n)=1$, and put

$$\|\xi\|_n = \int_{A_n} |x(\xi)|\,d\mu(x). \tag{17.1}$$

Theorem 17.1. The characteristic function $\chi(\xi)$ is continuous in the topology defined by $\{\|\cdot\|_n\}$.

Proof. For a given $\varepsilon>0$, choose n such that $\mu(A_n^c)<\varepsilon$. Then $\|\xi\|_n<\varepsilon$ implies that

$|1-\chi(\xi)| \leq \int_X |1-\exp(ix(\xi))|\,d\mu(x) \leq \int_{A_n} |x(\xi)|\,d\mu(x)+2\mu(A_n^c)<3\varepsilon$. (q.e.d.)

Remark. Th. 17.1 holds even if we replace $\|\cdot\|_n$ by $\|\cdot\|_{np}$

where

$$(17.2) \qquad \|\xi\|_{np} = \left(\int_{A_n} |x(\xi)|^p d\mu(x)\right)^{1/p}, \quad 1 \leq p < \infty$$

or by $\|\cdot\|_{A_n}$ in the meaning of (16.2).

Theorem 17.2. Let X be a separable metrizable locally convex space and Y be a complete metrizable vector space imbedded continuously in X. If there exists a Y-quasi-invariant Borel measure on X, then some neighbourhood of 0 in Y is precompact in X.

Proof. Since X is metrizable and separable, any Borel measure μ on X is concentrated on a σ-precompact set. (c f. Th.10.1 of Part A). Namely we have $^{\exists}K_n$, precompact and closed, $\lim_{n\to\infty}\mu(K_n)=1$.

Put $R=X^*$=the topological dual space of X, and imbed X into R'. Regarding μ as a measure on $(R', \mathcal{B}_R)$, we shall apply Th. 16.2 and Remark of Th. 17.1, then we see that $^{\exists}V$, neighbourhood of 0 in Y, $^{\exists}n$, $\overset{\circ}{V} \supset \overset{\circ}{K}_n$, hence $V \subset (\overset{\circ}{K}_n)^{\circ}$. Since $(\overset{\circ}{K}_n)^{\circ}$ is the symmetric convex hull of K_n, it is also precompact, therefore V is precompact in X. (q.e.d.)

Remark 1. The crucial point of the proof is that μ concentrates on $\bigcup_{n=1}^{\infty} K_n$. Using this, we can prove Th. 17.2 directly. Since $\mu(\bigcup_{n=1}^{\infty} K_n)=1$, evidently we have $Y_\mu \subset \bigcup_{n=1}^{\infty} K_n - \bigcup_{n=1}^{\infty} K_n = \bigcup_{n,m=1}^{\infty} (K_n - K_m)$. Thus Y_μ is σ-precompact. Since $Y \subset Y_\mu$ implies $Y \subset \bigcup_{n,m=1}^{\infty} (K_n - K_m)$, by Baire's theorem some $K_n - K_m$ has an inner point. Therefore some neighbourhood of 0 in Y is precompact in X.

Remark 2. Denote the Borel field of X with $\mathcal{B}$, then regarding X as a subspace of R' $(R=X^*)$, we have $\mathcal{B} = \mathcal{B}_R \cap X$.

Since every $\xi \in R$ is continuous on X, $\mathcal{B}_R \cap X \subset \mathcal{B}$ is evident.

Conversely, we shall show that every open set O belongs to $\mathcal{B}_R \cap X$. Since X is metrizable and separable, X satisfies Lindelöf's property (any open covering of an open set has a countable subcovering). So every weakly open set is $\mathcal{B}_R \cap X$-measurable, because it can be expressed as a countable union of weak neighbourhoods. Thus every weakly closed set is $\mathcal{B}_R \cap X$-measurable, hence particularly every closed ball in a continuous semi-norm is $\mathcal{B}_R \cap X$-measurable. Again by Lindelöf's property, every open set O in X can be expressed as a countable union of translations of such neighbourhoods, so O is $\mathcal{B}_R \cap X$-measurable.

<u>Corollary</u> of Th. 17.2. Let X be a complete separable metrizable locally convex space. (For instance let X be a separable Banach space or Frechet space). Then Y_μ does not contain any infinite dimensional closed subspace.

<u>Proof</u>. If Y is a closed subspace of X, then Y is complete in the topology of X. So $Y \subset Y_\mu$ implies that Y is locally precompact, therefore Y must be finite dimensional. (q.e.d.)

<u>Example 17.1</u>. (ℓ^p) is a separable Banach space for $1 \leq p < \infty$. If $p' < p$, we have $(\ell^{p'}) \subset (\ell^p)$ but the unit ball of $(\ell^{p'})$ is not precompact in (ℓ^p). Therefore by Th. 17.2, no Borel measure on (ℓ^p) is $(\ell^{p'})$-quasi-invariant.

(ℓ^∞) is a Banach space but not separable. Later in §27, we shall show the existence of (ℓ^1)-quasi-invariant measure on (ℓ^∞). So the assumption of separability is essential.

<u>Theorem 17.3</u>. Let X be a Hilbert space and Y be another Hilbert space imbedded continuously in X. There exists a Y-quasi-invariant measure on X if and only if the imbedding $Y \to X$ is of Hilbert-Schmidt type. (A measure is considered on the σ-

algebra $\mathcal{B}_{X^*}$, the smallest one in which all $\xi \in X^*$ are measurable).

Proof. The condition is sufficient, because using the inner product $(\, , \,)_Y$ of Y, we can define the corresponding Gaussian measure which is Y-quasi-invariant and lies on X. (c f. Th.9.1 and Th.8.2).

Conversely, if $Y \subset Y_\mu^{\bullet}$, then by Th. 16.2 the dual norm of Y is continuous in the characteristic topology τ_μ. But for any measure on X, the characteristic topology is weaker than Sazonov topology on X^* (§18 of Part A). Thus the dual inner product $(\, , \,)_Y'$ is of Hilbert-Schmidt type with respect to $(\, , \,)_X'$. Therefore $(\, , \,)_X$ is of Hilbert-Schmidt type with respect to $(\, , \,)_Y$. (c f. Lemma 22.1 in Part A). (q.e.d.)

Theorem 17.3 can be improved in the following way.

Theorem 17.3'. Let X be a Hilbert space and Y be a complete metrizable vector space imbedded continuously in X. There exists a Y-quasi-invariant measure on X if and only if $(\, , \,)_X$ is of Hilbert-Schmidt type with respect to some continuous inner product $(\, , \,)_H$ on Y.

Proof. To prove the sufficiency, consider the Gaussian measure corresponding to $(\, , \,)_H$.

For the necessity, we must modify the proof of Th. 17.3 as follows: From Th. 16.2 there exists a neighbourhood V of 0 in Y such that $\|\cdot\|_V$ is continuous in the Sazonov topology on X^*. This means that there exists a Hilbertian norm $\|\cdot\|_H'$ on X^* such that $\|\cdot\|_V \leq \|\cdot\|_H'$ and $(\, , \,)_H'$ is of Hilbert-Schmidt type with respect to $(\, , \,)_X'$. Then, the dual norm $\|\cdot\|_H$ of $\|\cdot\|_H'$ satisfies the requested conditions on Y. (q.e.d.)

We shall give here one more application using Th. 17.1.

Theorem 17.4. Let (a_n) be a sequence of positive numbers, and put

$$(17.3) \qquad (\ell^p)_a = \{(x_n)\in \mathbb{R}^\infty;\ \sum_{n=1}^{\infty}|a_n x_n|^p<\infty\},\ (1\leq p<\infty).$$

If $(\ell^{p'})_a$-quasi-invariant measure exists on (ℓ^p), then we must have $\sum_{n=1}^{\infty} a_n^{-p}<\infty$. (This theorem strengthens the result of Example 17.1).

Proof $p'\geq 1$ implies $(\ell^{p'})_a\supset(\ell^1)_a$, so that it is sufficient to prove the theorem for $p'=1$. Consider the semi-norms on $\mathbb{R}_0^\infty$ given by

$$(17.4) \qquad ||\xi||_n = \int_{nB}\Big|\sum_{j=1}^{\infty} x_j\xi_j\Big|d\mu(x),\ \xi\in\mathbb{R}_0^\infty.$$

Here B means the unit ball of (ℓ^p). From Th. 16.2 and Th. 17.1, if $Y_\mu\supset(\ell^1)_a$, then the dual norm $||\cdot||'$ of $(\ell^1)_a$ is continuous in the topology defined by $\{||\cdot||_n\}$. This means that $\exists n,\ \exists C>0$, $||\xi||'\leq C||\xi||_n$.

Consider $e_k=(0,0,\cdots,1,\cdots)$ (only the k-th coordinate is 1 and others are 0). Then evidently $||e_k||'=1/a_k$. So that we have

$$(17.5) \qquad 1/a_k\leq C\ ||e_k||_n = C\int_{nB}|x_k|d\mu(x).$$

Take the p-th power of both hand sides and apply the Hölder's inequality, then we get

$$(17.6) \qquad a_k^{-p}\leq C^p\int_{nB}|x_k|^p d\mu(x).$$

Summing up with respect to k, we get $\sum_{k=1}^{\infty} a_k^{-p}\leq C^p\int_{nB}\sum_{k=1}^{\infty}|x_k|^p d\mu(x)$ $\leq C^p n^p<\infty$. This proves the theorem. (q.e.d.)

§18. Kakutani topology

Another important tool used for the evaluation of Y_μ is Kakutani topology. Let $(X,\mathcal{B})$ be a measurable space. The Kakutani metric between two probability measures μ and ν is defined as:

$$(18.1)\qquad d(\mu,\nu) = \left(\int_X \left|\sqrt{\frac{d\mu}{d\lambda}} - \sqrt{\frac{d\nu}{d\lambda}}\right|^2 d\lambda\right)^{1/2} = \left\|\sqrt{\frac{d\mu}{d\lambda}} - \sqrt{\frac{d\nu}{d\lambda}}\right\|_{L^2(\lambda)},$$

where λ is a measure such that $\mu \lesssim \lambda$ and $\nu \lesssim \lambda$. The metric $d(\mu,\nu)$ does not depend on the choice of λ.

Another metric $d_1(\mu,\nu)$, called the metric of total variation, is defined as follows:

$$(18.2)\qquad d_1(\mu,\nu) = \int_X \left|\frac{d\mu}{d\lambda} - \frac{d\nu}{d\lambda}\right| d\lambda = \left\|\frac{d\mu}{d\lambda} - \frac{d\nu}{d\lambda}\right\|_{L^1(\lambda)}.$$

Since L^1-norm is the dual norm of L^∞, (18.2) can be written as

$$(18.3)\qquad d_1(\mu,\nu) = \sup_{\varphi \in S}\left(\int_X \varphi\, d\mu - \int_X \varphi\, d\nu\right),$$

where S is the set of all $\mathcal{B}$-measurable functions such that $\|\varphi\|_\infty = \sup_{x\in X} |\varphi(x)| \leqq 1$.

Two metrics d and d_1 are equivalent on the set of all probability measures, because we have

$$d_1(\mu,\nu) = \left\|\left(\sqrt{\frac{d\mu}{d\lambda}} - \sqrt{\frac{d\nu}{d\lambda}}\right)\left(\sqrt{\frac{d\mu}{d\lambda}} + \sqrt{\frac{d\nu}{d\lambda}}\right)\right\|_{L^1(\lambda)}$$

$$\leqq \left\|\sqrt{\frac{d\mu}{d\lambda}} - \sqrt{\frac{d\nu}{d\lambda}}\right\|_{L^2(\lambda)} \left\|\sqrt{\frac{d\mu}{d\lambda}} + \sqrt{\frac{d\nu}{d\lambda}}\right\|_{L^2(\lambda)} \leqq 2d(\mu,\nu), \text{ and}$$

$$d_1(\mu,\nu) \geqq \left\|\left(\sqrt{\frac{d\mu}{d\lambda}} - \sqrt{\frac{d\nu}{d\lambda}}\right)^2\right\|_{L^1(\lambda)} = \left\|\sqrt{\frac{d\mu}{d\lambda}} - \sqrt{\frac{d\nu}{d\lambda}}\right\|^2_{L^2(\lambda)} = d(\mu,\nu)^2.$$

Let R be a vector space, and μ be a probability measure on $(R',\mathcal{B}_R)$. Then we can define a metric d_μ on R' by

$$(18.4)\qquad d_\mu(x,y) = d(\mu_x,\mu_y),$$

the right hand side being in the sense of (18.1).

Definition 18.1. The metric d_μ in (18.4) defines a topology on R'. This topology is called Kakutani topology of μ.

The addition is continuous in Kakutani topology, because we have $d_\mu(x,y)=d_\mu(x-y,0)$. The latter comes from

$$(18.5) \qquad \left(\frac{d\mu}{d\lambda}\right)_y = \frac{d\mu_y}{d\lambda_y} \quad \text{where} \quad f_y(x) = f(x-y),$$

hence we get $d_\mu(x,y)^2=\int_{R'} \left|\sqrt{\frac{d\mu_x}{d\lambda}} - \sqrt{\frac{d\mu_y}{d\lambda}}\right|^2 d\lambda =$

$$\int_{R'} \left|\left(\sqrt{\frac{d\mu_{x-y}}{d\lambda_{-y}}}\right)_y - \left(\sqrt{\frac{d\mu}{d\lambda_{-y}}}\right)_y\right|^2 d\lambda = \int_{R'} \left|\sqrt{\frac{d\mu_{x-y}}{d\lambda_{-y}}} - \sqrt{\frac{d\mu}{d\lambda_{-y}}}\right|^2 d\lambda_{-y} = d_\mu(x-y,0)^2.$$

The scalar multiplication is, however, not necessarily continuous in Kakutani topology.

Kakutani topology is defined also by the metric

$$(18.4)' \qquad d_{1\mu}(x,y) = d_1(\mu_x,\mu_y),$$

the right hand side being in the sense of (18.2) or (18.3). This metric also satisfies $d_{1\mu}(x,y)=d_{1\mu}(x-y,0)$.

Theorem 18.1. Kakutani topology is stronger than the weak topology on R'.

Proof. It is sufficient to prove that every $\xi \in R$ is continuous in Kakutani topology, namely that

$$(18.6) \qquad \forall \xi \in R, \ \exists \delta>0, \ d_{1\mu}(x,0)<\delta \Longrightarrow |x(\xi)|<1.$$

But (18.6) is derived from

$$(18.7) \qquad \forall \xi \in R, \ \exists n, \ d_{1\mu}(x,0)<1/3 \Longrightarrow |x(\xi)|<n,$$

because $d_{1\mu}(x,0)<1/3n$ implies $d_{1\mu}(nx,0)<1/3$, and $|nx(\xi)|<n$ implies $|x(\xi)|<1$. We shall prove (18.7).

For every $A\in\mathcal{B}_R$, from (18.3) we see $d_{1\mu}(x,0)\geqq|\mu_x(A)-\mu(A)|$. If $\mu(A)>2/3$ and $d_{1\mu}(x,0)<1/3$, then $\mu_x(A)=\mu(A-x)>1/3$, so that $A\cap(A-x)\neq\phi$, hence $x\in A-A$. For a given ξ, put $A=\{x\in R';\ |x(\xi)|<n\}$, then we have $^\exists n$, $\mu(A)>2/3$. Then $x\in A-A$ means $|x(\xi)|<2n$.

(q.e.d.)

Theorem 18.2. $d_{1\mu}(x,0)$ is a lower semi-continuous function of x in the weak topology.

Proof. Since $\int\varphi\, d\mu_x=\int\varphi_{-x}\, d\mu$, using (18.3) and (18.4)' we have

$$d_{1\mu}(x,0) = \sup_{\varphi\in S}\int(\varphi_{-x}-\varphi)d\mu. \tag{18.8}$$

Put

$$C = \{\varphi(x)=f(x(\xi_1),\cdots,x(\xi_n));\ ^\exists\xi_1,\xi_2,\cdots,\xi_n\in R,\ ^\exists f \text{ continuous function of } n \text{ real variables with a compact support}\}. \tag{18.9}$$

Since $C\cap S$ is dense in S in $L^1(\lambda)$ where $\lambda=\frac{\mu_x+\mu}{2}$, $\sup\limits_{\varphi\in S}$ in (18.8) can be replaced by $\sup\limits_{\varphi\in C\cap S}$, thus we get

$$d_{1\mu}(x,0) = \sup_{\varphi\in C\cap S}\int(\varphi_{-x}-\varphi)d\mu. \tag{18.8)'$$

For each $\varphi\in C\cap S$, the integral $\int(\varphi_{-x}-\varphi)d\mu=\int(\varphi(y+x)-\varphi(y))d\mu(y)$ is evidently weakly continuous in x, so the supremum $d_{1\mu}(x,0)$ is lower semi-continuous in the weak topology. (q.e.d.)

Corollary 1. The closed ball in $d_{1\mu}$, i.e.

$$U_\varepsilon = \{x\in R';\ d_{1\mu}(x,0)\leqq\varepsilon\} \tag{18.10}$$

is weakly closed.

Corollary 2. If x_n tends to x weakly, then for every y we have $d_{1\mu}(x,y)\leqq\varliminf d_{1\mu}(x_n,y)$.

Theorem 18.3. R' is complete in Kakutani topology. Y_μ is a closed subset of R', hence Y_μ is also complete in Kakutani topology.

Proof. From Th. 18.1, every Cauchy sequence $\{x_n\}$ in Kakutani topology is a Cauchy sequence in the weak topology also, so x_n tends to some x weakly. (R' is weakly complete). From Corollary 2 of Th. 18.2, we have $d_{1\mu}(x,x_m)\leq\underline{\lim}_n\, d_{1\mu}(x_n,x_m)\to 0$ $(m\to\infty)$, so that x_m tends to x in Kakutani topology. Thus R' is complete.

For every $B\in\mathcal{B}_R$, $\mu(B-x)$ is a continuous function of x in Kakutani topology, because $d_{1\mu}(x,y)\geq|\mu(B-x)-\mu(B-y)|$, therefore $N_B=\{x\in R';\ \mu(B-x)=0\}$ is a closed set in Kakutani topology. Put $N=\bigcap_B\{N_B;\ \mu(B)=0\}$, then we have $Y_\mu=N\cap(-N)$ so that Y_μ is closed in Kakutani topology. (q.e.d.)

Denote with Y_μ^0 the largest vector space contained in Y_μ.

$$(18.11)\qquad Y_\mu^0 = \bigcap_{\alpha>0}(\alpha Y_\mu) = \{x\in R';\ \mu_{\alpha x}\sim\mu \quad \text{for } \forall\alpha\in\mathbb{R}\}.$$

Theorem 18.4. If $x\in Y_\mu^0$, then the map $\mathbb{R}\ni\alpha\to\alpha x\in R'$ is continuous in Kakutani topology, namely we have $\lim_{\alpha\to 0} d_{1\mu}(\alpha x,0)=0$.

Proof. Let ℓ_x be the line generated by x, and let m be the one-dimensional Lebesgue measure lying on ℓ_x. m can be regarded as a measure on $(R',\mathcal{B}_R)$.

Evidently m is invariant under the translation: $y\to y+\alpha x$, namely $m_{\alpha x}=m$ for $\forall\alpha\in\mathbb{R}$. Consider the measure $\lambda=m*\mu$, then we have $\lambda_{\alpha x}=\lambda$ for $\forall\alpha\in\mathbb{R}$. Since $x\in Y_\mu^0$ and m lies on ℓ_x, we have $\lambda\sim\mu$, hence by Radon-Nikodym's theorem we have $d\mu=fd\lambda$ for some $f\in L^1(\lambda)$.

First we shall assume that $f\in C$ (in the meaning of (18.9)), then $d_{1\mu}(\alpha x,0)=\int_{R'}|f(y+\alpha x)-f(y)|d\lambda$ satisfies evidently $\lim_{\alpha\to 0} d_{1\mu}(\alpha x,0)=0$. Next, for general $f\in L^1(\lambda)$, we can choose some $g\in C$ such that $\|f-g\|_{L^1(\lambda)}<\varepsilon$. Then

$$d_{1\mu}(\alpha x,0)=\int_{R'}|f(y+\alpha x)-f(y)|d\lambda\leqq 2\int_{R'}|f(y)-g(y)|d\lambda$$
$$+\int_{R'}|g(y+\alpha x)-g(y)|d\lambda<2\varepsilon+\int_{R'}|g(y+\alpha x)-g(y)|d\lambda.$$

So for sufficiently small α we have $d_{1\mu}(\alpha x,0)<3\varepsilon$. (q.e.d.)

Theorem 18.5. Let Y be a subspace of R' with a complete metrizable vector topology stronger than the weak topology. If $Y\subset Y_\mu$, then the topology of Y is stronger than Kakutani topology.

Corollary. If Y is a finite dimensional subspace contained in Y_μ, then Kakutani topology is identical with the Euclid topology on Y.

For the proof, combine Th. 18.5 with Th. 18.1.

Proof of Th. 18.5. Since U_ε in (18.10) is weakly closed, $U_\varepsilon\cap Y$ is closed in Y. From Th. 18.4, we have $\forall y\in Y$, $\exists n$, $\frac{1}{n}y\in U_\varepsilon$ so that $Y=\bigcup_{n=1}^{\infty}(nU_\varepsilon\cap Y)$. So by Baire's theorem, some $nU_\varepsilon\cap Y$ has an inner point in Y. Namely $\exists y_0\in Y$, $\exists V$ neighbourhood of 0 in Y, $y_0+V\subset nU_\varepsilon$, therefore $\frac{1}{n}V\subset U_\varepsilon-U_\varepsilon\subset U_{2\varepsilon}$. This proves that the topology of Y is stronger than Kakutani topology on Y.

(q.e.d.)

Owing to Th. 18.4, we can linearize Kakutani topology on the space Y_μ^0 (c.f. §14). The linearized Kakutani topology is defined by the metric

$$(18.12)\qquad \bar{d}_\mu(x,y)=\sup_{0\leqq\alpha\leqq 1} d_\mu(\alpha x,\alpha y),\ \text{or}$$

$$(18.12)'\qquad \bar{d}_{1\mu}(x,y)=\sup_{0\leqq\alpha\leqq 1} d_{1\mu}(\alpha x,\alpha y).$$

Remark. Th. 18.5 holds even if we replace Kakutani topology by the linearized Kakutani topology.

Theorem 18.6. Y_μ^0 is complete in the linearized Kakutani topology.

Proof. Suppose that $\{x_n\}$ is a Cauchy sequence in the linearized Kakutani topology on Y_μ^0, then $\{\alpha x_n\}$ is a Cauchy sequence in Kakutani topology for $0 \leqq \alpha \leqq 1$, so that $\exists y_\alpha \in Y_\mu$, $d_\mu(y_\alpha, \alpha x_n) \to 0$ $(n \to \infty)$. This convergence is uniform in α, so if we can show $\exists y \in Y_\mu^0$, $y_\alpha = \alpha y$, then the proof will be completed.

Since αx_n tends to y_α in Kakutani topology, y_α is the weak limit of αx_n, (by Th. 18.1). Especially y_1 is the weak limit of x_n, and we have $y_\alpha = \alpha y_1$. Put $A = \{\alpha \in \mathbb{R};\ \alpha y_1 \in Y_\mu\}$, then since $\alpha y_1 = y_\alpha \in Y_\mu$ for $0 \leqq \alpha \leqq 1$, we have $A \supset [0,1]$. But A being an additive group, we must have $A = \mathbb{R}$. This means $y_1 \in Y_\mu^0$.

(q.e.d.)

Theorem 18.7. The linearized Kakutani topology is the unique vector topology on Y_μ^0 which is complete, metrizable and stronger than the weak topology.

Corollary. If $Y_\mu^0 = Y_\nu^0$, then the linearized Kakutani topology of μ is identical with that of ν.

Proof of Th. 18.7. Let τ be a complete metrizable vector topology on Y_μ^0 stronger than the weak topology. By Th. 18.5 and its Remark, τ is stronger than the linearized Kakutani topology. So it turns out that two topologies are identical because of Borel's theorem:

Let E and F be two complete metrizable vector spaces and f be an algebraical isomorphism from E onto F. If f is continuous, then f^{-1} is also continuous, hence f becomes a homeomorphism.

(q.e.d.)

§19. Evaluation of Y_μ in terms of Kakutani topology

Let μ be a probability measure on $(R',\mathcal{B}_R)$. We shall define a function $F(x)$ on Y_μ^0 as

$$(19.1) \qquad F(x) = \int_{R'} \sqrt{\frac{d\mu_x}{d\mu}}\, d\mu, \quad x \in Y_\mu^0.$$

Theorem 19.1. $F(x)$ is positive definite on Y_μ^0, and $F(\alpha x)$ is continuous in α for every $x \in Y_\mu^0$.

Proof. Using (18.5), we have $F(x)=\int_{R'} \sqrt{\frac{d\mu_{x+y}}{d\mu_y}}\, d\mu_y$ $=\int_{R'} \sqrt{\frac{d\mu_{x+y}}{d\mu}} \sqrt{\frac{d\mu_y}{d\mu}}\, d\mu$, hence

$$(19.2) \qquad F(x-y) = \int_{R'} \sqrt{\frac{d\mu_x}{d\mu}} \sqrt{\frac{d\mu_y}{d\mu}}\, d\mu.$$

Therefore we get $\sum_{j,k=1}^{n} \alpha_j \bar{\alpha}_k F(x_j - x_k) = \int_{R'} \left| \sum_{j=1}^{n} \alpha_j \sqrt{\frac{d\mu_{x_j}}{d\mu}} \right|^2 d\mu \geq 0.$
This shows that $F(x)$ is positive definite on Y_μ^0.

From the definition of Kakutani metric, we have

$$(19.3) \qquad d_\mu(x,0)^2 = \int_{R'} \left| \sqrt{\frac{d\mu_x}{d\mu}} - 1 \right|^2 d\mu = 2(1-F(x)), \quad x \in Y_\mu^0.$$

So by Th. 18.4, $F(\alpha x)$ is a continuous function of α for $\forall x \in Y_\mu^0$. (q.e.d.)

Definition 19.1. Owing to Th. 19.1, we can consider a measure $\tilde{\mu}$ on $(Y_\mu^{0\prime}, \mathcal{B}_{Y_\mu^0})$ whose characteristic function is $F(x)$. Namely

$$(19.4) \qquad \int_{Y_\mu^{0\prime}} \exp(if(x))\, d\tilde{\mu}(f) = F(x), \quad \forall x \in Y_\mu^0.$$

The measure $\tilde{\mu}$ is called the adjoint measure of μ.

Remark. The correspondence $\mu \to \tilde{\mu}$ is neither one-to-one, nor onto.

Theorem 19.2. The linearized Kakutani topology on Y_μ^0 is nothing but the characteristic topology of $\tilde{\mu}$.

Proof. From (19.3), Kakutani topology is the weakest one in

which $F(x)$ is continuous, so the linearized Kakutani topology is the weakest vector topology in which $F(x)$ is continuous. Hence we get the theorem. (q.e.d.)

Theorem 19.3. Let Y be a subspace of R' with a complete metrizable vector topology stronger than the weak topology. Let Y_0 be a dense subspace of Y, and μ be a Y_0-quasi-invariant measure on $(R', \mathcal{B}_R)$. Then μ is Y-quasi-invariant if and only if $F(x)$, defined on Y_0, is continuous in the topology of Y.

Proof. If $Y_\mu \supset Y$, then by Th. 18.5 the topology of Y is stronger than Kakutani topology, so $F(x)$ is continuous in the topology of Y.

Conversely, assume that $F(x)$ is continuous in the topology of Y. Since Y_0 is dense in Y, for every $y \in Y$ there exists a sequence $\{x_n\} \subset Y_0$ such that $x_n \to y$ in Y. Since $\{x_n\}$ is a Cauchy sequence in the topology of Y, under the assumption $\{x_n\}$ is a Cauchy sequence in Kakutani topology also. Therefore x_n tends to some $x \in Y_\mu$ in Kakutani topology. Since both the topology of Y and Kakutani topology are stronger than the weak topology, both x and y are the weak limit of x_n, thus we get $y = x \in Y_\mu$. (q.e.d.)

Now, we shall give some applications of Th. 19.3.

Let $\mathbb{R}^\infty$ be the infinite product of real lines. It is the algebraical dual of $\mathbb{R}_0^\infty = \{(\xi_n);\ ^\exists N,\ \xi_n = 0 \text{ for } n \geq N\}$. Then $\mathcal{B}_{\mathbb{R}_0^\infty}$ is the weak Borel field of $\mathbb{R}^\infty$.

Theorem 19.4. Let μ be a weak Borel measure on $\mathbb{R}^\infty$. If μ is $\mathbb{R}_0^\infty$-quasi-invariant (regarding $\mathbb{R}_0^\infty$ as a subspace of $\mathbb{R}^\infty$), then Y_μ is properly larger than $\mathbb{R}_0^\infty$. More precisely, we have

(19.5) $\quad Y_\mu \supset \mathbb{R}_0^\infty \Rightarrow {}^\exists a=(a_n),\ Y_\mu \supset H_a,$

(19.6) $\quad H_a = \{(x_n)\in\mathbb{R}^\infty;\ \sum_{n=1}^\infty a_n^2 x_n^2 < \infty\}.$

Proof. Since $Y_\mu \supset \mathbb{R}_0^\infty$, we can consider the adjoint measure $\tilde{\mu}$ on $\mathbb{R}^\infty$. But for every probability measure ν on $\mathbb{R}^\infty$, we have ${}^\exists b =(b_n)$, $b_n>0, \nu(H_b)=1$ as proved below: For each n we have ${}^\exists c_n>0$, $\nu(|x_n|>c_n)<2^{-n}$, so that for ν-almost all $x=(x_n)$, we have $\overline{\lim}\ \frac{|x_n|}{c_n} \leq 1$, hence $\sum_{n=1}^\infty \frac{x_n^2}{n^2 c_n^2} < \infty$. Thus putting $b_n=1/nc_n$, we have $\nu(H_b)=1$.

Suppose that $\tilde{\mu}(H_b)=1$. Then the characteristic topology of $\tilde{\mu}$, namely the linearized Kakutani topology of μ, is weaker than the dual norm of H_b. The topological dual space of H_b is H_a where $a_n=1/b_n$, and $\mathbb{R}_0^\infty$ is evidently dense in H_a in the dual norm, so by Th. 19.3 we have $Y_\mu \supset H_a$. (q.e.d.)

Corollary. Let μ be an $\mathbb{R}_0^\infty$-quasi-invariant measure on $\mathbb{R}^\infty$. Then there exists an $\mathbb{R}_0^\infty$-quasi-invariant Gaussian measure g such that $\mu \sim g * \mu$.

Proof. Suppose that $Y_\mu \supset H_a$. Put $\alpha_n = na_n$, then H_α is imbedded in H_a in a Hilbert-Schmidt way. Let g be the Gaussian measure corresponding to the norm of H_α, then g is H_α-quasi-invariant and lies on H_a. Since $Y_\mu \supset H_a$ and g lies on H_a, we have $\mu \sim g * \mu$. (q.e.d.)

Using this corollary, we have an interesting result on the absolute continuity of two $\mathbb{R}_0^\infty$-quasi-invariant measures.

Theorem 19.5. Let μ and ν be $\mathbb{R}_0^\infty$-quasi-invariant probability measures on $\mathbb{R}^\infty$. Put

(19.7) $\quad \mathcal{B}_{inv} = \{B\in\mathcal{B};\ B+\xi=B \ \text{ for } {}^\forall\xi\in\mathbb{R}_0^\infty\},$

where $\mathcal{B}$ is the weak Borel field of $\mathbb{R}^\infty$. Then we have

(1) $\quad \nu \lesssim \mu \Longleftrightarrow \nu|\mathcal{B}_{inv} \lesssim \mu|\mathcal{B}_{inv}$

(2) $\quad \mu$ is $\mathbb{R}_0^\infty$-ergodic $\Longleftrightarrow {}^\forall B \in \mathcal{B}_{inv}$, $\mu(B)=0$ or 1.

Here $\mu|\mathcal{B}_{inv}$ means the restriction of μ on $\mathcal{B}_{inv}$.

Proof $\Rightarrow$ part is evident. We shall prove the $\Longleftarrow$ part.

Consider an $\mathbb{R}_0^\infty$-quasi-invariant Gaussian measure g such that $\mu \sim g * \mu$ and $\nu \sim g * \nu$. Then, for ${}^\forall B \in \mathcal{B}$ we have

$$\text{(19.8)} \qquad \mu(B) = 0 \Longleftrightarrow (g*\mu)(B) = 0 \Longleftrightarrow \int_{\mathbb{R}^\infty} g(B-x)\,d\mu(x) = 0$$

$$\Longleftrightarrow \mu(\{x \in \mathbb{R}^\infty;\ g(B-x)>0\}) = 0.$$

The same holds for the measure ν. Since g is $\mathbb{R}_0^\infty$-quasi-invariant, the set $\{x \in \mathbb{R}^\infty;\ g(B-x)>0\}$ belongs to $\mathcal{B}_{inv}$. So from (19.8) we see that $\nu|\mathcal{B}_{inv} \lesssim \mu|\mathcal{B}_{inv}$ implies $\nu \lesssim \mu$ on $\mathcal{B}$.

Next, we shall prove the $\Leftarrow$ part of (2). Suppose that both μ and ν are $\mathbb{R}_0^\infty$-quasi-invariant. Then $\nu \lesssim \mu$ implies $\nu|\mathcal{B}_{inv} \lesssim \mu|\mathcal{B}_{inv}$, hence under the assumption $\mu|\mathcal{B}_{inv} = 0$ or 1, we have $\nu|\mathcal{B}_{inv} = \mu|\mathcal{B}_{inv}$, so that from (1) we see $\nu \sim \mu$. This means that μ is $\mathbb{R}_0^\infty$-ergodic on $\mathcal{B}$. (q.e.d.)

Corollary. Let μ be an $\mathbb{R}_0^\infty$-quasi-invariant measure on $\mathbb{R}^\infty$. Put

$$\text{(19.9)} \qquad \mu_n(B) = \mu(\mathbb{R}^n \times B), \quad {}^\forall B \in \mathcal{B}.$$

Then we have $\mu \sim m \times \mu_n$, where m is the n-dimensional Lebesgue measure.

Proof. Let m' be a probability measure equivalent with m. Then both μ and $m' \times \mu_n$ are $\mathbb{R}_0^\infty$-quasi-invariant, and $\mu(\mathbb{R}^n \times B)$

$=\mu_n(B)=(m'\times\mu_n)(\mathbb{R}^n\times B)$. Since every set in $\mathcal{B}_{inv}$ is written in the form of $\mathbb{R}^n\times B$, this implies $\mu|\mathcal{B}_{inv}=(m'\times\mu_n)|\mathcal{B}_{inv}$, hence we have $\mu\sim m'\times\mu_n\sim m\times\mu_n$. (q.e.d.)

We shall give another example of the evaluation of Y_μ.

Theorem 19.6. No Borel measure on $\mathbb{R}^\infty$ satisfies $Y_\mu=(\ell^p)$ for $p>2$. (In other words, if $Y_\mu\supset(\ell^p)$, then Y_μ should be properly larger than (ℓ^p)).

Remark. For $p\leq 2$, we can find a measure such that $Y_\mu=(\ell^p)$. See Examples 2 and 3 in §23.

Proof. Assume $Y_\mu=(\ell^p)$, then by Th. 18.7, the topology of (ℓ^p) is identical with the linearized Kakutani topology of μ, hence with the characteristic topology of $\tilde{\mu}$. So this theorem is derived from the corollary of the following Th. 19.7.

Definition 19.2. Let E be a normed space, and m be the infinite product of uniform probability measures on $[-1,1]$. E is said to be of cotype 2, if

$$(19.10)\qquad \forall(\xi_n)_{n=1}^\infty\subset E,\ \sup_N\int\Big\|\sum_{i=1}^N a_i\xi_i\Big\|^2 dm(a)<\infty\Rightarrow\sum_{i=1}^\infty\|\xi_i\|^2<\infty.$$

Remark 1. The definition is equivalent even if we take m as the infinite product of $(\delta_1+\delta_{-1})/2$ where δ_1 and δ_{-1} are the Dirac measures placed at 1 and -1 respectively.

Remark 2. (ℓ^p) is not of cotype 2 for $p>2$. Because, putting $\xi_i=\alpha_i e_i$, we have $\sum_{i=1}^\infty\|\xi_i\|^2=\sum_{i=1}^\infty|\alpha_i|^2$ and $\int\|\sum_{i=1}^N a_i\xi_i\|^2\ dm(a)=\int(\sum_{i=1}^N|a_i\alpha_i|^p)^{2/p}\ dm(a)\leq(\sum_{i=1}^N|\alpha_i|^p)^{2/p}$. But if $p>2$, there exists a sequence (α_i) which satisfies $\sum_{i=1}^\infty|\alpha_i|^p<\infty$ but $\sum_{i=1}^\infty|\alpha_i|^2=\infty$.

Theorem 19.7. Let $(X,\mathcal{B},\mu)$ be a probability measure space. If the space $L^0(\mu)$ (=the space of all $\mathcal{B}$-measurable functions with the topology of measure convergence) has a subspace isomorphic with a normed space, then this norm should be of cotype 2.

Corollary. If the characteristic topology is a normed topology, then this norm should be of cotype 2.

Proof. Let E be a normed subspace of $L^0(\mu)$, and we shall prove (19.10). We shall divide the proof into two parts. Namely

$$\sup_N \int \| \sum_{i=1}^N a_i \xi_i \|^2 dm(a) < \infty \Rightarrow \sum_{i=1}^\infty |\xi_i(x)|^2 < \infty \text{ for } \mu\text{-almost all } x. \tag{19.11}$$

$$\sum_{i=1}^\infty |\xi_i(x)|^2 < \infty \text{ for } \mu\text{-almost all } x \Rightarrow \sum_{i=1}^\infty \|\xi_i\|^2 < \infty. \tag{19.12}$$

First we shall prove (19.12). Suppose $\sum_{i=1}^\infty |\xi_i(x)|^2 < \infty$ for μ-almost all x, then putting $d\nu = \exp(-\sum_{i=1}^\infty |\xi_i(x)|^2) d\mu$ we have $\mu \sim \nu$. Thus the topology of measure convergence of μ is identical with that of ν. Since $L^2(\nu)$ is imbedded continuously in $L^0(\nu)$, we have

$$\exists c>0, \|\xi\| \leq c \|\xi\|_{L^2(\nu)} \quad \text{for } \forall \xi \in E \cap L^2(\nu). \tag{19.13}$$

Therefore we have $\sum_{i=1}^\infty \|\xi_i\|^2 \leq c^2 \sum_{i=1}^\infty \|\xi_i\|^2_{L^2(\nu)} = c^2 \int_X \sum_{i=1}^\infty \xi_i(x)^2 \times \exp(-\sum_{i=1}^\infty \xi_i(x)^2) d\mu(x) < \infty$. This proves (19.12).

Next, we shall prove (19.11). Since E is a subspace of $L^0(\mu)$, for any $\varepsilon>0$ there exists $\delta>0$ such that $\int_X \cos(\xi(x)) \times d\mu(x) \geq 1-\varepsilon$ for $\|\xi\| \leq \delta$ and ≥ -1 otherwise. Thus we get

$$\int_X \cos(\xi(x)) d\mu(x) \geq 1-\varepsilon-\frac{2}{\delta^2}\|\xi\|^2, \quad \forall \xi \in E. \tag{19.14}$$

From (19.14) we have

$$\iint_X \cos(\sum_{j=1}^N a_j \xi_j(x)) d\mu(x) dm(a) \geq 1-\varepsilon-\frac{2M}{\delta^2}, \text{ where} \tag{19.15}$$

$$M = \sup_N \int \| \sum_{j=1}^N a_j \xi_j \|^2 dm(a). \tag{19.16}$$

The left hand side of (19.15) is equal with $\int_X \prod_{j=1}^{N} \frac{\sin\xi_j(x)}{\xi_j(x)} d\mu(x)$,

because $\int \cos(\sum_{j=1}^{N} a_j t_j) dm(a) = \mathrm{Re}\int \exp(i\sum_{j=1}^{N} a_j t_j) dm(a) =$

$\prod_{j=1}^{N} \frac{1}{2}\int_{-1}^{1} \exp(iat_j) da = \prod_{j=1}^{N} \frac{\sin t_j}{t_j}$. Consider the absolute value of the integrand and take the limit of $N\to\infty$, then we have

$$(19.17) \qquad \int_X \prod_{j=1}^{\infty} \left|\frac{\sin\xi_j(x)}{\xi_j(x)}\right| d\mu(x) \geqq 1-\varepsilon-\frac{2M}{\delta^2}.$$

The integrand of (19.17)=0 if $\sum_{j=1}^{\infty}|\xi_j(x)|^2=\infty$, because $\frac{\sin t}{t} \sim 1-\frac{t^2}{6}$ for $t\approx 0$. Thus we get

$$(19.18) \qquad \mu(A) \geqq 1-\varepsilon-\frac{2M}{\delta^2} \quad \text{where}$$

$$(19.19) \qquad A = \{x\in X;\ \sum_{j=1}^{\infty} |\xi_j(x)|^2<\infty\}.$$

Now, we shall replace ξ_j with $c\xi_j$ for a constant c, then M is replaced with c^2M. Since $\sum_{j=1}^{\infty} |\xi_j(x)|^2<\infty$ is equivalent with $\sum_{j=1}^{\infty} |c\xi_j(x)|^2<\infty$, we have $\mu(A)\geqq 1-\varepsilon-\frac{2c^2M}{\delta^2}$.

Letting $c\to 0$, we get $\mu(A)\geqq 1-\varepsilon$. Since $\varepsilon>0$ is arbitrary, letting $\varepsilon\to 0$ we have $\mu(A)\geqq 1$. (q.e.d.)

CHAPTER 4. PRODUCT MEASURES ON $\mathbb{R}^\infty$

§20. Product of one-dimensional probability measures

In Chapter 4, we shall restrict our consideration to the product measures on $\mathbb{R}^\infty$, and discuss the quasi-invariance, especially with respect to (ℓ^2), of such measures.

Let μ_k be a probability Borel measure on $\mathbb{R}^1$, and consider their product μ on $\mathbb{R}^\infty$.

$$(20.1) \qquad \mu = \prod_{k=1}^{\infty} \mu_k$$

Theorem 20.1. The measure μ above is $\mathbb{R}_0^\infty$-quasi-invariant, if and only if each μ_k is equivalent with the Lebesgue measure, namely if and only if

$$(20.2) \qquad \forall k,\ \exists f_k > 0,\ d\mu_k = f_k(x_k)dx_k.$$

Under this condition, μ is $\mathbb{R}_0^\infty$-ergodic.

Proof. Suppose that μ is $\mathbb{R}_0^\infty$-quasi-invariant, then for the projection $p^{(k)}: x=(x_1,x_2,\cdots) \to x_k$, $p^{(k)}\circ\mu=\mu_k$ is $\mathbb{R}^1$-quasi-invariant, so that μ_k is equivalent with the Lebesgue measure (Th. 1.1 of Chapter 1).

Conversely, suppose that each μ_k is $\mathbb{R}^1$-quasi-invariant. Then for $x=(x_1,x_2,\cdots,x_n,0,\cdots)\in\mathbb{R}^n\times\{0\}$, we have $\tau_x\mu= \prod_{k=1}^{n}(\tau_{x_k}\mu_k)\times\prod_{k=n+1}^{\infty}\mu_k$. Since $\tau_{x_k}\mu_k \sim \mu_k$, we get $\tau_x\mu \sim \mu$. Combining this with $\bigcup_{n=1}^{\infty}(\mathbb{R}^n\times\{0\})=\mathbb{R}_0^\infty$, we get the $\mathbb{R}_0^\infty$-quasi-invariance of μ.

In order to show that μ is $\mathbb{R}_0^\infty$-ergodic, we shall prove that every bounded measurable function $\varphi(x)$ satisfying $\forall y \in \mathbb{R}_0^\infty$, $\forall' x$, $\varphi(x)=\varphi(x+y)$ should be constant almost everywhere. (cf. (E5) in §6)

Let M_n be the subspace of $L^2(\mu)$ which consists of all functions depending only on $x_1, x_2, \cdots, x_n$. M_n is a closed subspace of $L^2(\mu)$, and $\bigcup_{n=1}^{\infty} M_n$ is dense in $L^2(\mu)$. Consider the orthogonal projection:

$$\varphi(x) = \varphi_1^{(n)}(x) + \varphi_2^{(n)}(x), \quad \varphi_1^{(n)} \in M_n, \quad \varphi_2^{(n)} \in M_n^\perp. \tag{20.3}$$

For $y = (y_1, y_2, \cdots, y_n, 0, \cdots) \in \mathbb{R}^n \times \{0\}$, we have $\varphi_1^{(n)}(x+y) \in M_n$ and $\varphi_2^{(n)}(x+y) \in M_n^\perp$. The former is evident. To prove the latter, consider $\psi \in M_n$, then

$$\int \psi(x)\varphi_2^{(n)}(x+y)d\mu(x) = \int \psi(x-y)\varphi_2^{(n)}(x)d(\tau_y\mu)(x)$$

$$= \int \psi(x-y) \prod_{k=1}^{n} \frac{d(\tau_{y_k}\mu_k)}{d\mu_k}(x_k)\varphi_2^{(n)}(x)d\mu(x) = 0.$$

From the uniqueness of the orthogonal projection, for every $y \in \mathbb{R}^n \times \{0\}$, $\varphi(x)=\varphi(x+y)$ implies $\varphi_1^{(n)}(x)=\varphi_1^{(n)}(x+y)$ and $\varphi_2^{(n)}(x)=\varphi_2^{(n)}(x+y)$. Since $\prod_{k=1}^{n} \mu_k$ is $\mathbb{R}^n$-ergodic, $\varphi_1^{(n)}$ should be constant. Thus, $\lim_{n\to\infty} \varphi_1^{(n)} = \varphi$ implies that φ should be constant. (q.e.d.)

Hereafter we shall assume (20.2). Let $\mathcal{B}_n$ be the smallest σ-algebra in which projections $p^{(k)}$; $1 \le k \le n$ are measurable. In other words we have $\mathcal{B}_n = \mathcal{B}_1^n \times \{\mathbb{R}^\infty, \phi\}$, where $\mathcal{B}_1$ is the Borel field of $\mathbb{R}^1$, $\mathcal{B}_1^n$ is the n-ple product of $\mathcal{B}_1$, and $\{\mathbb{R}^\infty, \phi\}$ is the σ-algebra generated only by $\mathbb{R}^\infty$ and ϕ.

From Th. 7.1 and its corollary, we have $y \in Y_\mu$, if and only if $\left\langle \sqrt{\frac{d(\tau_y\mu|\mathcal{B}_n)}{d(\mu|\mathcal{B}_n)}}, \sqrt{\frac{d(\tau_y\mu|\mathcal{B}_m)}{d(\mu|\mathcal{B}_m)}} \right\rangle_{L^2(\mu)}$ tends to 1 as $n, m \to \infty$. Substituting (20.2) into this, we have $y \in Y_\mu$ if and only if

$$\prod_{k=n+1}^{m} \int \sqrt{f_k(x_k)f_k(x_k-y_k)}\, dx_k \xrightarrow[n,m\to\infty]{} 1. \tag{20.4}$$

But (20.4) is equivalent with

$$\prod_{k=1}^{\infty} \int \sqrt{f_k(x_k)f_k(x_k-y_k)}\, dx_k > 0. \tag{20.4)'$$

Theorem 20.2. If μ is in the form of (20.1) and (20.2), then we have $y \in Y_\mu$ if and only if (20.4)' holds.

Denote the Fourier transform of $\sqrt{f_k(x_k)}$ with $g_k(z_k)$, then the adjoint measure $\tilde{\mu}$ is given on $\mathbb{R}^\infty$ by $d\tilde{\mu}(z) = \prod_{k=1}^{\infty} |g_k(z_k)|^2 dz_k$. Using this, we have $y \in Y_\mu$ if and only if

$$\prod_{k=1}^{\infty} \int \cos(y_k z_k)|g_k(z_k)|^2 dz_k > 0. \tag{20.5}$$

Proof. First half has been proved already.

Since $\mathcal{F}(\sqrt{f_k}) = g_k$ implies $\mathcal{F}(\sqrt{f_k(x_k-y_k)}) = \exp(iz_k y_k)g_k(z_k)$, we have

$$\int \sqrt{f_k(x_k)f_k(x_k-y_k)}\, dx_k = \int \exp(iz_k y_k)|g_k(z_k)|^2 dz_k,$$

but $\exp(iz_k y_k)$ in the right hand side can be replaced by

$\cos(z_k y_k)$, because $|g_k(z_k)|^2$ is an even function. (For the definition of the adjoint measure, cf. Definition 19.1).

(q.e.d.)

We shall define R_μ by

$$(20.6) \qquad R_\mu = \{y=(y_k);\ \sum_{k=1}^{N} y_k x_k \text{ converges in } L^0(\mu) \text{ as } N\to\infty\}.$$

For $y \in R_\mu$, we shall denote this limit with $\sum_{k=1}^{\infty} y_k x_k$. Evidently R_μ becomes a linear subspace of $\mathbb{R}^\infty$.

Remark 1. If μ is a product measure (20.1), then $\sum_{k=1}^{\infty} y_k x_k$ converges in $L^0(\mu)$ if and only if it converges almost everywhere.

Theorem 20.3. If μ is in the form of (20.1) and (20.2), then we have $Y_\mu^0 = R_{\tilde\mu}$.

Remark 2. Since $|g_k(z)|^2$ is an even function, $\sum_{k=1}^{\infty} y_k z_k$ converges in $L^0(\tilde\mu)$ if and only if $\sum_{k=1}^{\infty} y_k^2 z_k^2$ converges in $L^0(\tilde\mu)$, hence if and only if $\sum_{k=1}^{\infty} y_k^2 z_k^2 < \infty$ almost everywhere. Therefore $y \in R_{\tilde\mu}$ is equivalent with $\tilde\mu(\{z \in \mathbb{R}^\infty;\ \sum_{k=1}^{\infty} y_k^2 z_k^2 < \infty\}) = 1$.

For the proof of Remark 1 and 2, see for instance M. Loeve; "Probability Theory". 4th edition, §17 and §18.

Proof (of Th.20.3). Since Y_μ^0 is complete in the linearized Kakutani topology (Th.18.6), and since the linearized Kakutani topology of μ is identical with the characteristic topology of $\tilde\mu$ (Th.19.2), $y \in R_{\tilde\mu}$ implies $y \in Y_\mu^0$. Because $y \in R_{\tilde\mu}$ means that $\{(y_1, y_2, \cdots, y_n, 0, \cdots)\}_{n=1,2,\cdots}$ becomes a Cauchy sequence in the linearized Kakutani topology, so that it converges to some $y' \in Y_\mu^0$ in the linearized Kakutani topology, hence in the weak topology also. Thus we get $y=y' \in Y_\mu^0$.

Conversely assume that $y \in Y_\mu^0$. Then we have $\lim_{\alpha \to 0}(\alpha y)=0$ in Kakutani topology (Th. 18.4), so that for sufficiently small α, say for $|\alpha| \leq \alpha_0$, we have

$$\prod_{k=1}^{\infty} \int \cos(\alpha y_k z_k)|g_k(z_k)|^2 dz_k > \frac{1}{2}, \text{ hence we have}$$

$$\sum_{k=1}^{\infty} \int \{1-\cos(\alpha y_k z_k)\}|g_k(z_k)|^2 dz_k < \log 2.$$

Averaging with α for $|\alpha| \leq \alpha_0$, we get

$$\sum_{k=1}^{\infty} \int \{1 - \frac{\sin(\alpha_0 y_k z_k)}{\alpha_0 y_k z_k}\} \mid g_k(z_k)|^2 dz_k < \log 2.$$

Therefore

$$\sum_{k=1}^{\infty} \{1 - \frac{\sin(\alpha_0 y_k z_k)}{\alpha_0 y_k z_k}\} < \infty \quad \text{for } \tilde{\mu}\text{-almost all } z, \text{ hence}$$

$$\sum_{k=1}^{\infty} y_k^2 z_k^2 < \infty \quad \text{for } \tilde{\mu}\text{-almost all } z \text{ because } 1-\frac{\sin t}{t} \sim \frac{t^2}{6}$$

for $t \fallingdotseq 0$. Thus from Remark 2, we get $y \in R_{\tilde{\mu}}$. (q.e.d.)

<u>Theorem 20.4</u>. Let Y be a linear subspace of $\mathbb{R}^\infty$ with a complete metrizable vector topology stronger than the weak topology. If $Y \subset R_\mu$, then the map $y \to \sum_{k=1}^{\infty} y_k x_k$ is continuous from Y into $L^0(\mu)$.

<u>Proof.</u> Since the topology of Y is stronger than the weak topology, the map $y \to \sum_{k=1}^{N} y_k x_k$ is continuous from Y into $L^0(\mu)$, so that for a closed neighbourhood V of 0 in $L^0(\mu)$,

the set $U_V=\{y\in Y;\ \forall N,\ \sum_{k=1}^{N} y_k x_k \in V\}$ is closed in Y.

We shall prove that $\bigcup_{n=1}^{\infty}(nU_V)=Y$. We shall choose a neighbourhood W of 0 in $L^0(\mu)$ such that $W+W\subset V$ and $\alpha W\subset W$ for $|\alpha|\leq 1$. Consider an arbitrary $y\in Y$. Since $L^0(\mu)$ is a topological vector space, we have $\forall N,\ \exists n_N$, $\frac{1}{n_N}\sum_{k=1}^{N} y_k x_k \in W$, and $\exists n_\infty$, $\frac{1}{n_\infty}\sum_{k=1}^{\infty} y_k x_k \in W$. Also we have $\exists N_0$, $N\geq N_0 \Longrightarrow \sum_{k=1}^{N} y_k x_k \in \sum_{k=1}^{\infty} y_k x_k + W$, so that $\frac{1}{n_\infty}\sum_{k=1}^{N} y_k x_k \in W+\frac{1}{n_\infty}W\subset V$. Thus, putting $n=\mathrm{Max}(n_1,n_2,\cdots,n_{N_0},n_\infty)$, we get $\frac{y}{n}\in U_V$, hence $y\in nU_V$.

Since Y is complete metrizable, by Baire's theorem some nU_V has an inner point in Y. Namely, $\exists y_0\in Y$, $\exists U$ (neighbourhood of 0 in Y), $y_0+U\subset nU_V$, therefore $\frac{1}{n}U\subset U_V+U_V\subset U_{V+V}$. This means that $y\in\frac{1}{n}U$ implies $\sum_{k=1}^{\infty} y_k x_k \in V+V$. This holds for any V, so that the map $y\to\sum_{k=1}^{\infty} y_k x_k$ is continuous from Y into $L^0(\mu)$. (q.e.d.)

<u>Remark 3</u>. If $R_\mu \supset (\ell^p)$, then we have $\forall y=(y_k)\in(\ell^p)$, $\sum_{k=1}^{\infty} y_k^2 x_k^2 < \infty$ for μ-almost all x, and the map $y\to\sum_{k=1}^{\infty} y_k^2 x_k^2$ is continuous from (ℓ^p) into $L^0(\mu)$.

Because, in the norm of (ℓ^p) we have $\|ay\|\leq\|y\|$ for any $a=(a_k)\in[0,1]^\infty$ where $ay=(a_k y_k)$, so that (19.11) assures that $\sum_{k=1}^{\infty} y_k^2 x_k^2 < \infty$ for μ-almost all x. The continuity of the map $y\to\sum_{k=1}^{\infty} y_k^2 x_k^2$ is proved in a quite similar way as Th. 20.4.

§21. Stationary product measures

Definition 21.1. In (20.2), if all f_k is identical with some function f, then the product measure μ:

$$\text{(21.1)} \qquad \mu = \mu_1^{\infty}, \quad d\mu_1 = f(x)dx$$

is said to be a stationary product measure. (In (21.1), of course we assume that $f(x) > 0$ and $\int_{-\infty}^{\infty} f(x)dx=1$).

Remark. In the terminology of the probability theory, (21.1) means that x_k follows the identical independent distribution μ_1.

Theorem 21.1. Let μ be a stationary product measure. Then we have

1) $Y_\mu \subset (\ell^2)$,
2) $R_\mu \subset (\ell^2)$, and
3) the characteristic topology of μ is stronger than the topology of (ℓ^2).

Proof. Let g be the Fourier transform of $\sqrt{f}$ and h be the Fourier transform of $|g|^2$. Since $|g|^2$ is an integrable function, we see that i) $h(t)$ is continuous, ii) $\lim_{t\to\pm\infty} h(t)=0$ and iii) $h(t)\neq 1$ for $t\neq 0$.

From Th. 20.2, $y=(y_k)\in Y_\mu$ implies that $\sum_{k=1}^{\infty}(1-h(y_k)) < \infty$. From this we get $\lim_{k\to\infty} h(y_k)=1$, so that $\lim_{k\to\infty} y_k=0$. (From ii) $\{y_k\}$ should be bounded, and from i) and iii) $\{y_k\}$ does not have any non-zero limitting point, thus $\lim_{k\to\infty} y_k=0$). But we have $\lim_{y\to 0} \frac{1-h(y)}{y^2} > 0$, because

$$\frac{1-h(y)}{y^2} = \int_{-\infty}^{\infty} \frac{1-\cos(yz)}{y^2}|g(z)|^2 dz \geq \frac{4}{\pi^2}\int_{-\pi/2y}^{\pi/2y} z^2|g(z)|^2 dz.$$

(The last inequality comes from $1-\cos t \geq \frac{4}{\pi^2}t^2$ for $|t| \leq \frac{\pi}{2}$).
Hence $\sum_{k=1}^{\infty}(1-h(y_k)) < \infty$ implies $\sum_{k=1}^{\infty} y_k^2 < \infty$, thus $y \in (\ell^2)$.
This assures $Y_\mu \subset (\ell^2)$, so we have completed the proof of 1).

From the above proof, we see that $\forall\varepsilon, \exists\delta$, $\sum_{k=1}^{\infty}(1-h(y_k))<\delta$ implies $\sum_{k=1}^{\infty} y_k^2 < \varepsilon$. This means that the characteristic topology of $\tilde{\mu}$ is stronger than the topology of (ℓ^2). If we assume that f is an even function, then putting $h=\mathcal{F}(f)$ (instead of $h=\mathcal{F}(|g|^2)$), we can show that the characteristic topology of μ is stronger than the topology of (ℓ^2).

For a general f, consider $\varphi=f*f$ (or $\varphi(x) = \int_{-\infty}^{\infty} f(t)f(t-x)dt$), then φ is an even function. The stationary product of φdx is equal with $\mu*\check{\mu}$, and the characteristic topology of μ is stronger than that of $\mu*\check{\mu}$. Therefore the characteristic topology of μ is stronger than the topology of (ℓ^2). Thus 3) has been proved.

If $y=(y_k)\in R_\mu$, then $\{(y_1,y_2,\cdots,y_n,0,\cdots)\}_{n=1,2,\cdots}$ becomes a Cauchy sequence in the characteristic topology of μ, hence in (ℓ^2) also, thus it has a limit y' in (ℓ^2). Since y is the weak limit of this sequence, we have $y=y'\in (\ell^2)$. This assures $R_\mu \subset (\ell^2)$, so we have completed the proof of 2). (q.e.d.)

<u>Theorem 21.2.</u> Let μ be the stationary product of fdx.
1) We have $Y_\mu=(\ell^2)$ if and only if $\frac{d}{dx}\sqrt{f} \in L^2(dx)$. Here $\frac{d}{dx}$ means the derivative in the sense of distribution.
2) We have $R_\mu=(\ell^2)$ if and only if $x\sqrt{f(x)}\in L^2(dx)$ and $\int_{-\infty}^{\infty} xf(x)dx=0$. Furthermore, then and only then the characteristic topology of μ is identical with the topology of (ℓ^2).

Proof. 1) Let g be the Fourier transform of $\sqrt{f}$, then $\frac{d}{dx}\sqrt{f}\in L^2(dx)$ is equivalent with $zg(z)\in L^2(dz)$. In this case, putting $I=\int_{-\infty}^{\infty} z^2|g(z)|^2dz$, we have $\int\sum_{k=1}^{\infty} y_k^2z_k^2d\tilde{\mu}(z)=\sum_{k=1}^{\infty} y_k^2 I$. So that whenever $y=(y_k)\in(\ell^2)$, we have $\sum_{k=1}^{\infty} y_k^2z_k^2<\infty$ for $\tilde{\mu}$-almost all z. Therefore from Th. 20.3 and Remark 2 of §20, we have $y\in R_{\tilde{\mu}}=Y_\mu^0$. This assures $(\ell^2)\subset Y_\mu$.

Conversely assume that $Y_\mu=(\ell^2)$. From Th. 20.4, the map $y=(y_k)\in(\ell^2)\to\sum_{k=1}^{\infty} y_kz_k\in L^0(\tilde{\mu})$ is continuous, so that

$$\exists\delta>0,\ \sum_{k=1}^{\infty} y_k^2\leq\delta^2\Longrightarrow\prod_{k=1}^{\infty}\int_{-\infty}^{\infty}\cos(y_kz)|g(z)|^2dz\geq\frac{1}{2}.$$

Now put $y_1=y_2=\cdots=y_n=\frac{\delta}{\sqrt{n}}$ and $y_k=0$ for $k>n$, then we have

$$\left\{\int_{-\infty}^{\infty}\cos\frac{\delta z}{\sqrt{n}}|g(z)|^2dz\right\}^n\geq\frac{1}{2},\ \text{hence}\ n\int_{-\infty}^{\infty}\left(1-\cos\frac{\delta z}{\sqrt{n}}\right)|g(z)|^2dz\leq\log 2.$$

Therefore, combining $\lim_{n\to\infty} n(1-\cos\frac{t}{\sqrt{n}})=\frac{t^2}{2}$ with Fatou's inequality $(\int\underline{\lim}\, f_n(x)dx\leq\underline{\lim}\int f_n(x)dx$ for $f_n\geq 0)$, we get $\frac{\delta^2}{2}\int_{-\infty}^{\infty} z^2|g(z)|^2dz\leq\log 2$. This means that $zg(z)\in L^2(dz)$.

2) Second half. Suppose that the topology of (ℓ^2) is stronger than the characteristic topology of μ. Then for every $y=(y_k)\in(\ell^2)$, $\{(y_1,y_2,\cdots,y_n,0,\cdots)\}_{n=1,2,\cdots}$ is a Cauchy sequence in the characteristic topology, hence in the topology of $L^0(\mu)$, so that $y\in R_\mu$. This assures $(\ell^2)\subset R_\mu$.

Conversely if $R_\mu=(\ell^2)$, then from Th. 20.4, the map $y=(y_k)\in(\ell^2)\to\sum_{k=1}^{\infty} y_kx_k\in L^0(\mu)$ is continuous, so that the topology of (ℓ^2) is stronger than the characteristic topology of μ.

<u>First half</u>. Assume that $x\sqrt{f(x)}\in L^2(dx)$ and $\int_{-\infty}^{\infty} xf(x)dx=0$. Since $1-\cos t \leq \frac{t^2}{2}$ and $|t-\sin t| \leq \frac{t^2}{2}$, putting $\int_{-\infty}^{\infty} x^2f(x)dx=I$, we have

$$|\int_{-\infty}^{\infty} (1-\exp(iyx))f(x)dx| \leq \frac{y^2}{\sqrt{2}} I.$$

Combining this with $|\prod_{k=1}^{\infty}(1+\alpha_k)-1| \leq \prod_{k=1}^{\infty}(1+|\alpha_k|)-1 \leq \exp(\sum_{k=1}^{\infty}|\alpha_k|)-1$, we have

$$|\int \exp(i\sum_{k=1}^{\infty} y_k x_k)d\mu(x)-1| \leq \prod_{k=1}^{\infty}(1+\frac{y_k^2}{\sqrt{2}}I)-1 \leq \exp(\frac{I}{\sqrt{2}}\sum_{k=1}^{\infty} y_k^2)-1,$$

therefore the topology of (ℓ^2) is stronger than the characteristic topology of μ.

Conversely assume that the topology of (ℓ^2) is stronger than the characteristic topology of μ. Then the former is stronger than the characteristic topology of $\mu * \check{\mu}$, which is a stationary product of an even measure. So, from the second half of the proof of 1) (using $f*\check{f}$ instead of $|g(z)|^2$), we see that $\int_{-\infty}^{\infty} x^2(f*\check{f})(x)dx=\int_{-\infty}^{\infty} x^2f(t)f(t-x)dtdx < \infty$, from which we can deduce $\int_{-\infty}^{\infty} x^2f(x)dx < \infty$.

Put $\int_{-\infty}^{\infty} xf(x)dx=a$, then $f_1(x)=f(x+a)$ satisfies both $\int_{-\infty}^{\infty} x^2f_1(x)dx < \infty$ and $\int_{-\infty}^{\infty} xf_1(x)dx=0$. Therefore the characteristic function of the stationary product of f_1dx is continuous in the topology of (ℓ^2), but it is written as $\exp(-ia\sum_{k=1}^{\infty} y_k)\chi(y)$, χ being the characteristic function of μ. Hence if and only if $a=0$, χ is continuous in the topology of (ℓ^2). (If $a\neq 0$, putting $y_1=y_2=\cdots=y_n=\frac{\pi}{na}$ and $y_k=0$ for $k>n$, we have $\exp(-ia\sum_{k=1}^{\infty} y_k)=-1$ but $\sum_{k=1}^{\infty} y_k^2=\pi^2/na^2 \to 0$

$(n \to \infty))$. (q.e.d.)

Example 1. $f(x) = C\exp(-|x|^r)$, where $r > 0$ and C is the normalization constant.

i) $x\sqrt{f(x)} \in L^2(dx)$ is evident, so that $R_\mu = (\ell^2)$ and the characteristic function of μ is continuous in (ℓ^2).

ii) $\frac{d}{dx}\sqrt{f(x)} = -\frac{r\sqrt{C}}{2}|x|^{r-1}\exp(-\frac{1}{2}|x|^r)$ implies that $Y_\mu = (\ell^2)$ for $r > \frac{1}{2}$ but $Y_\mu \subsetneqq (\ell^2)$ for $r \leq \frac{1}{2}$.

Example 2. $f(x) = C(1+x^2)^{-r}$, where $r > \frac{1}{2}$ and C is the normalization constant.

i) $x\sqrt{f(x)} \sim x^{-r+1}$ for $x \to \infty$ implies that we have $R_\mu = (\ell^2)$ if and only if $r > \frac{3}{2}$. Then and only then the characteristic function of μ is continuous in (ℓ^2).

ii) $\frac{d}{dx}\sqrt{f(x)} = -\sqrt{C}rx(1+x^2)^{-\frac{r}{2}-1} \sim x^{-r-1}$ for $x \to \infty$ implies that $Y_\mu = (\ell^2)$ for all $r > \frac{1}{2}$.

§22. Gaussian measures and stationary products

Let $(\ ,\ )_1$ be an inner product on $\mathbb{R}_0^\infty$. If it induces the topology of (ℓ^2), (namely if $\|\cdot\|_1$ is equivalent with the usual (ℓ^2)-norm), then the corresponding gaussian measure is i) (ℓ^2)-quasi-invariant and ii) the characteristic topology is identical with the topology of (ℓ^2).

On the other hand, many stationary product measures satisfy both i) and ii), for instance, the case of $f(x) = C\exp(-|x|^r)$ for $r > \frac{1}{2}$ (Example 1 of §21) and the case of $f(x) = C(1+x^2)^{-r}$ for $r > \frac{3}{2}$ (Example 2). However, they are essentially different from gaussian measures, as proved below.

In Chapter 2, we have considered gaussian measures with zero mean. If we include gaussian measures with non-zero mean, the characteristic function is given by

$$\chi(\xi) = \exp(iy(\xi) - \tfrac{1}{2}\|\xi\|_1^2) \tag{22.1}$$

for some $y \in \mathbb{R}^\infty$ and some inner product $(\ ,\ )_1$.

<u>Theorem 22.1.</u> If the stationary product measure of $f dx$ is equivalent with some gaussian measure, then f itself should be gaussian, namely we must have

$$f(x) = \frac{1}{\sqrt{2\pi}c} \exp\left(-\frac{(x-m)^2}{2c^2}\right) \tag{22.2}$$

for some constant m and c.

<u>Remark 1</u>. The stationary product of (22.2) is evidently a gaussian measure, with $y=(m,m,\cdots)$ and $(\ ,\ )_1 = c(\ ,\ )$ where $(\ ,\ )$ is the usual (ℓ^2)-inner product.

<u>Proof</u>. <u>First Step</u>. Suppose that the stationary product measure μ is equivalent with a gaussian measure whose characteristic function is given by (22.1). Then $\|\cdot\|_1$ should induce the topology of (ℓ^2), as proved below.

Since the characteristic topology of μ is stronger than the topology of (ℓ^2), the topology induced by $\|\cdot\|_1$ should be stronger than (ℓ^2). Since $Y_\mu \subset (\ell^2)$, the topological dual of $\mathbb{R}_0^\infty$ in the semi-norm $\|\cdot\|_1$ is contained in (ℓ^2). This means that the topology induced by $\|\cdot\|_1$ is weaker than (ℓ^2).

<u>Second Step</u>. Consider a linear operator A on $\mathbb{R}_0^\infty$ such that

$$(\xi,\eta)_1 = (A\xi, A\eta) \quad \text{for } \forall \xi,\eta \in \mathbb{R}_0^\infty. \tag{22.3}$$

Then the gaussian measure corresponding to (22.1) is given by $\tau_y(A^*\circ g)$, where g is the usual gaussian measure corresponding to $(\ , \ )$.

If $\tau_y(A^*\circ g) \sim \tau_{m1}(cI\circ g)$, where $m1=(m,m,\cdots)$ and I is the identity operator, then we have $\mu \sim \tau_{m1}(cI\circ g)$. This implies (22.2), because the stationary products of f_1dx and f_2dx are mutually singular if $f_1 \neq f_2$. (Example of §7).

Therefore it is sufficient to prove $\tau_y(A^*\circ g) \sim \tau_{m1}(cI\circ g)$. For this purpose, from §10, it is sufficient to show that

(22.4) 1) $y-m1 \in (\ell^2)$ and
2) c^2I-A^*A is of Hilbert-Schmidt type.

<u>Third Step</u>. Since μ is a stationary product measure, it is invariant under any permutation $\sigma: (x_k) \to (x_{\sigma(k)})$. So, $\tau_y(A^*\circ g)$ is quasi-invariant under the permutation σ. Since $\sigma\circ(\tau_y(A^*\circ g)) = \tau_{\sigma y}(\sigma A^*\circ g)$, this implies

(22.5) $\forall\sigma$(permutation $(x_k) \to (x_{\sigma(k)})$),
1) $y-\sigma y \in (\ell^2)$ and
2) $I-A^{*-1}\sigma A^*A\sigma A^{-1}$ is of Hilbert-Schmidt.

Now we shall deduce (22.4) from (22.5).

<u>Lemma 1</u>. If $y=(y_k)$ satisfies $y-\sigma y \in (\ell^2)$ for every permutation σ, then we have $y-m1 \in (\ell^2)$ for some constant m.

<u>Proof.</u> First remark that if σ is a one-to-one map from $\mathbb{N}_1$ to $\mathbb{N}_2$, where both $\mathbb{N}_1$ and $\mathbb{N}_2$ are subsets of $\mathbb{N}$ (=the set of all natural numbers) such that $\mathbb{N}_1^c$ and $\mathbb{N}_2^c$ are infinite sets, then σ can be extended to a permutation on $\mathbb{N}$.

Now we shall prove that $\lim_{k\to\infty} y_k$ exists. If $\{y_k\}$ is not a Cauchy sequence, then we have $^{\exists}\delta > 0$, $^{\exists}\{n_k\}$, $^{\forall}k$, $|y_k - y_{n_k}| \geq \delta$. Then putting $\sigma(k)=n_k$ for even k and defining $\sigma(k)$ suitably for odd k, we see that $y-\sigma y \notin (\ell^2)$. This contradicts to the assumption.

Next, putting $z_k = y_k - \lim_{k\to\infty} y_k$, we shall show $(z_k) \in (\ell^2)$. Excluding zero terms from the first, we can assume that $^{\forall}k$, $z_k \neq 0$. Since $\lim_{k\to\infty} z_k = 0$, we have $^{\exists}\{n_k\}$, $^{\forall}k$, $|z_k - z_{n_k}| \geq \frac{1}{2}|z_k|$. Putting $\sigma(k)=n_k$ for even k and defining $\sigma(k)$ suitably for odd k, we see that $z-\sigma z \in (\ell^2)$ implies $\sum_{k=1}^{\infty} z_{2k}^2 < \infty$. In the same way, we can show $\sum_{k=1}^{\infty} z_{2k+1}^2 < \infty$. Thus we have $\sum_{k=1}^{\infty} z_k^2 < \infty$, hence $z \in (\ell^2)$. (q.e.d.)

<u>Lemma 2</u>. If a linear bounded operator B on (ℓ^2) satisfies the condition that $B-\sigma B\sigma$ is of Hilbert-Schmidt for every permutation σ, then for some constant c, $cI-B$ is of Hilbert-Schmidt.

<u>Remark 2</u>. If we apply Lemma 2 to (22.5) as $B=A^*A$, then 2) of (22.5) implies 2) of (22.4).

<u>Proof.</u> We shall express B as a matrix (B_{jk}), then $\sigma B\sigma$ corresponds to the matrix $(B_{\sigma(j)\sigma(k)})$. So that $B-\sigma B\sigma$ is of Hilbert-Schmidt if and only if $\sum_{j,k=1}^{\infty} |B_{jk} - B_{\sigma(j)\sigma(k)}|^2 < \infty$. From this we shall deduce $\sum_{j,k=1}^{\infty} |B_{jk} - c\delta_{jk}|^2 < \infty$.

From Lemma 1, we see $\sum_{j=1}^{\infty} |B_{jj} - c|^2 < \infty$ for some c.

So it is sufficient to show $\sum_{j\neq k} B_{jk}^2 < \infty$, namely $\sum_{j=1}^{\infty} S_j < \infty$ where $S_j = (\sum_{k=1}^{j-1} + \sum_{k=j+1}^{\infty}) B_{jk}^2$. Since B is a bounded operator, we see $S_j < \infty$ for every j.

We shall remark $|\sqrt{S_{\sigma(j)}}-\sqrt{S_j}|^2 \leq \sum_{k=1}^{\infty} |B_{jk}-B_{\sigma(j)\sigma(k)}|^2$, so applying Lemma 1 to $\{\sqrt{S_j}\}$, we have $\sum_{j=1}^{\infty} \sqrt{S_j}^2 = \sum_{j=1}^{\infty} S_j < \infty$ if we prove $\lim_{j\to\infty} S_j = 0$.

We shall prove $\lim_{j\to\infty} S_j = 0$. Put $R_{jN} = \sum_{k=N}^{\infty} B_{jk}^2$ for $N > j$ and $S_{jN} = \sum_{k=1}^{N} B_{jk}^2$ for $N < j$. Since B and B^* are bounded, we have $\lim_{N\to\infty} R_{jN} = \lim_{j\to\infty} S_{jN} = 0$. So that for a given j, if we choose N and J sufficiently large, we get $\sqrt{S_j - R_{jN}} - \sqrt{S_{JN}} \geq \frac{1}{2}\sqrt{S_j}$.

Suppose that σ maps $\{1,2,\cdots,j-1,j+1,\cdots,N\}$ to $\{1,2,\cdots,N-1\}$, and that $\sigma(j)=J$, then we have

$$(22.6) \qquad \sum_{k=1}^{\infty} |B_{jk}-B_{\sigma(j)\sigma(k)}|^2 \geq \sum_{k=1}^{N} |B_{jk}-B_{J\sigma(k)}|^2$$

$$\geq \left|\sqrt{S_j - R_{jN}} - \sqrt{S_{JN}}\right|^2 \geq \tfrac{1}{4} S_j .$$

If $\overline{\lim_{j\to\infty}} S_j = \alpha$, then there exists a subsequence $\{j_p\}$ such that $\lim_{p\to\infty} S_{j_p} = \alpha$. Taking a suitable subsequence of $\{j_p\}$ if necessary, we can construct a permutation σ which assures (22.6) for $j=j_p$, $p=1,2,\cdots$. Then summing up with p, we get $\sum_{p=1}^{\infty} S_{j_p} < \infty$ which implies $\lim_{p\to\infty} S_{j_p} = \overline{\lim_{j\to\infty}} S_j = 0$.

(q.e.d.)

§23. Estimation of Y_μ and R_μ

In Th. 21.2, we have stated the condition for $Y_\mu = (\ell^2)$ or $R_\mu = (\ell^2)$ for a stationary product measure μ. This section is devoted to complementary discussions for some cases not satisfying this condition. Throughout this section, we shall assume that $f(x)$ is an even function.

Theorem 23.1. Assume that $f(x)$ is an even function and satisfies $\exists r > 1$, $\lim_{x\to\infty} x^r f(x)=c > 0$. Then the stationary product μ of fdx satisfies $R_\mu=(\ell^p)$ and the characteristic topology of μ is identical with the topology of (ℓ^p). Here, $p=2$ for $r > 3$, $p=r-1$ for $1<r<3$ and $p=2-$ for $r=3$, where (ℓ^{2-}) means

$$(23.1) \qquad (\ell^{2-}) = \{y=(y_k)\in \mathbb{R}^\infty;\ \sum_{k=1}^{\infty} y_k^2(1+|\log|y_k||) < \infty\}.$$

Example 1. If $f(x)=C(1+x^2)^{-r}$, then $R_\mu=(\ell^{2r-1})$ for $\frac{1}{2}<r<\frac{3}{2}$ and $R_\mu=(\ell^{2-})$ for $r=\frac{3}{2}$. (c.f. Example 2 of §21).

Proof. If $r > 3$, then we have $x\sqrt{f(x)}\in L^2(dx)$ so that $R_\mu = (\ell^2)$ by Th. 21.2.

Assume $1<r\leq 3$. Since the characteristic function is given by $\chi(\xi)= \prod_{k=1}^{\infty}\int_{-\infty}^{\infty} \cos(\xi_k x)f(x)dx$ for $\xi=(\xi_k)\in \mathbb{R}_0^\infty$, we have $1-\chi(\xi) \sim \sum_{k=1}^{\infty} |\xi_k|^p$ if we prove

$$(23.2) \qquad \int_{-\infty}^{\infty} (1-\cos(ax))f(x)dx \sim a^p.$$

Here $\sim$ means that the both hand sides have the same infinitesimal order.

To prove (23.2), we shall rewrite the left hand side into $I_1(a)+I_2(a)$ where

$$I_1(a) = 2\int_0^\alpha (1-\cos(ax))f(x)dx \leq a^2\int_0^\alpha x^2 f(x)dx \quad \text{and}$$

$$I_2(a) = 2\int_\alpha^\infty (1-\cos(ax))f(x)dx = 2\int_{a\alpha}^\infty (1-\cos t)f(\tfrac{t}{a})\frac{dt}{a}$$

$$= 2a^{r-1}\int_{a\alpha}^{\infty}\frac{1-\cos t}{t^r}(\frac{t}{a})^r f(\frac{t}{a})dt.$$

We shall choose α such that $x^r f(x)$ is bounded for $x > \alpha$. Then, by Lebesgue's theorem we have

$$\lim_{a\to 0} I_2(a)/a^{r-1} = 2c\int_0^{\infty}\frac{1-\cos t}{t^r}\,dt \quad \text{if} \quad r < 3.$$

If $r=3$, then $\frac{1-\cos t}{t^3}$ is not integrable, but $\frac{1-\cos t}{t^3} \sim \frac{1}{t}$ for $t \to 0$ implies $I_2(a) \sim a^2|\log a|$.

Thus we have proved (23.2), so that the characteristic topology of μ is identical with the topology of (ℓ^p).

From this, we get $R_\mu = (\ell^p)$ in a similar way with the proof of 2) of Th. 21.2. (q.e.d.)

<u>Theorem 23.2</u>. Assume that $f(x)$ is an even function, twice continuously differentiable for $x > 0$ and that $x^r\frac{d^2}{dx^2}\sqrt{f(x)}$ is bounded and has non-zero limit as $x \to 0$. Then the stationary product μ of fdx satisfies $Y_\mu = (\ell^p)$ where $p=5-2r$ for $\frac{3}{2} < r < \frac{5}{2}$ and $p=2-$ for $r=\frac{3}{2}$.

<u>Example 2.</u> If $f(x) = C\exp(-|x|^r)$, then we have

$$\frac{d^2}{dx^2}\sqrt{f(x)} = \sqrt{C}\{-\frac{r(r-1)}{2} + \frac{r^2}{4}x^r\}x^{r-2}\exp(-\tfrac{1}{2}x^r),$$ so that $Y_\mu = (\ell^{2r+1})$ for $0 < r < \frac{1}{2}$ and $Y_\mu = (\ell^{2-})$ for $r=\frac{1}{2}$. (c.f. Example 1 of §21).

<u>Proof</u>. From Th. 20.2, we have $y=(y_k)\in Y_\mu$ if and only if $\prod_{k=1}^{\infty}\int_{-\infty}^{\infty}\cos(y_k z)|g(z)|^2dz > 0$. Therefore applying Th. 23.1 to $\tilde{\mu}$, we get $Y_\mu = (\ell^p)$ if we prove that $\lim_{z\to\infty} z^{3-r}g(z)\neq 0$ exists.

Denote $\sqrt{f(x)}$ with $\varphi(x)$. From the assumption we see that $\lim_{x\to\infty}\varphi(x)=\lim_{x\to\infty}\varphi'(x)=0$ and $\lim_{x\to 0}x\varphi(x)=\lim_{x\to 0}x^2\varphi'(x)=0$. So calculating the integration by parts, we have

$$g(z) = \sqrt{\frac{2}{\pi}}\int_0^\infty \cos(xz)\varphi(x)dx = -\sqrt{\frac{2}{\pi}}\frac{1}{z}\int_0^\infty \sin(xz)\varphi'(x)dx$$

$$= \sqrt{\frac{2}{\pi}}\frac{1}{z^2}\int_0^\infty (1-\cos(xz))\varphi''(x)dx.$$

If we change the integral variable to $t=xz$ we have

$$g(z) = \sqrt{\frac{2}{\pi}}\frac{1}{z^3}\int_0^\infty (1-\cos t)\varphi''(\tfrac{t}{z})dt$$

$$= \sqrt{\frac{2}{\pi}}\frac{1}{z^{3-r}}\int_0^\infty \frac{1-\cos t}{t^r}\,(\tfrac{t}{z})^r\varphi''(\tfrac{t}{z})dt.$$

The integral in the right hand side tends to $c\int_0^\infty \frac{1-\cos t}{t^r}\,dt$ as $z\to\infty$, where $c=\lim_{x\to 0}x^r\varphi''(x)$. Thus we get $g(z)\sim z^{r-3}$ as $z\to\infty$. (q.e.d.)

Example 3. If $f(x)=|x|^r$ on a neighbourhood of 0, $f(x)=\exp(-|x|)$ for sufficiently large $|x|$, and twice continuously differentiable for $x>0$, then the stationary product μ of $f dx$ satisfies $Y_\mu=(\ell^{r+1})$ for $-1<r<1$. (For $r=0$, consider $(\log|x|)^2$ instead of $|x|^0$).

Because, we have $\frac{d^2}{dx^2}x^{r/2}=\frac{r(r-2)}{4}\,x^{(r-4)/2}$, so applying Th. 23.2 we get $p=5-2\cdot\frac{4-r}{2}=r+1$.

§24. Non-stationary product measures

Now, we shall consider non-stationary product measures. Namely, consider a measure μ on $\mathbb{R}^\infty$ such that

$$(24.1)\qquad d\mu = \prod_{k=1}^{\infty} f_k(x_k)dx_k,$$

$$f_k(x) > 0, \int_{-\infty}^{\infty} f_k(x)dx = 1.$$

We shall generalize Th. 21.1 in this section, and Th. 21.2 in the next section.

Theorem 24.1. If $\{\sqrt{f_k(x)}\}_{k=1}^{\infty}$ is precompact in $L^2(dx)$, then we have 1) $Y_\mu \subset (\ell^2)$, 2) $R_\mu \subset (\ell^2)$, and 3) the characteristic topology of μ is stronger than the topology of (ℓ^2).

Remark. $\{\sqrt{f_k(x)}\}_{k=1}^{\infty}$ is precompact in $L^2(dx)$, if and only if $\{f_k(x)\}_{k=1}^{\infty}$ is precompact in $L^1(dx)$. (c.f. §18, equivalence of d and d_1).

Proof. First Step. We shall prove that

$$(24.2)\qquad \operatorname*{Inf}_{k}\ \operatorname*{Inf}_{a>0}\ \mathrm{Max}(1, \tfrac{1}{a^2})\int_{-\infty}^{\infty}(1-\cos(ax))f_k(x)dx > 0.$$

If (24.2) does not hold, we would have

$$(24.3)\qquad \forall n,\ \exists k_n,\ \exists a_n > 0,\ \int_{-\infty}^{\infty}(1-\cos(a_n x))f_{k_n}(x)dx \leq \tfrac{1}{n}\mathrm{Min}(1,a_n^2).$$

Since $\{f_{k_n}\}$ is precompact in $L^1(dx)$, choosing a suitable subsequence if necessary, we can assume that $f_{k_n} \to \exists f_\infty$ in $L^1(dx)$. Then from (24.3) we have $\lim_{n\to\infty}\int_{-\infty}^{\infty}(1-\cos(a_n x))f_\infty(x)dx =0$, so that $\{a_n\}$ should tend to zero. Again from (24.3), we get

$$\int_{-\infty}^{\infty}\frac{1-\cos(a_n x)}{a_n^2}\, f_{k_n}(x)dx \leq \frac{1}{n}$$

so that taking the limit of $n \to \infty$, we have $\int_{-\infty}^{\infty} x^2 f_\infty(x)dx=0$, which is impossible.

Second Step. Let $g_k(z)$ be the Fourier transform of $\sqrt{f_k(x)}$. Then, $\{g_k\}_{k=1}^{\infty}$ is precompact in $L^2(dz)$, therefore $\{|g_k(z)|^2\}_{k=1}^{\infty}$ is precompact in $L^1(dz)$. From Th. 20.2, we have

$$(24.4) \qquad y \in Y_\mu \Longrightarrow \sum_{k=1}^{\infty} \int_{-\infty}^{\infty} (1-\cos(y_k z)) |g_k(z)|^2 dz < \infty.$$

Apply (24.2) replacing $f_k(x)$ by $|g_k(z)|^2$, then (24.4) implies that $\sum_{k=1}^{\infty} \mathrm{Min}(1, y_k^2) < \infty$, so we have $y=(y_k) \in (\ell^2)$. This proves $Y_\mu \subset (\ell^2)$.

The above procedure also proves 3), when every $f_k(x)$ is an even function. If $f_k(x)$ is not even, then $\varphi_k = f_k * \check{f}_k$ is even and $\{\varphi_k\}_{k=1}^{\infty}$ is precompact in $L^1(dx)$ (because $\|\varphi_j - \varphi_k\|_{L^1} \leq 2\|f_j - f_k\|_{L^1}$). The product of $\varphi_k dx_k$ is equal with $\mu * \check{\mu}$, whose characteristic topology is weaker than that of μ. Therefore the characteristic topology of μ is stronger than the topology of (ℓ^2).

If $y \in R_\mu$, then $\{(y_1, y_2, \cdots, y_n, 0, \cdots)\}_{n=1}^{\infty}$ is a Cauchy sequence in the characteristic topology, so in the topology of (ℓ^2) also. This means $y \in (\ell^2)$. (q.e.d.)

Corollary. If the product measure μ given in (24.1) satisfies either of 1) $Y_\mu \supset (\ell^p)$, 2) $R_\mu \supset (\ell^p)$ or 3) the characteristic function is continuous in (ℓ^p) for some $p > 2$, then $\{\sqrt{f_k(x)}\}_{k=1}^{\infty}$ has no limit point in $L^2(dx)$.

Because if some subsequence $\{\sqrt{f_{k_n}(x)}\}_{n=1}^{\infty}$ tends to $\sqrt{f_\infty(x)}$ in $L^2(dx)$, then by Th. 24.1 we have $Y_\mu \subset \{(y_k);\ \sum_{n=1}^{\infty} y_{k_n}^2 < \infty\}$, which contradicts to $Y_\mu \supset (\ell^p)$, $p > 2$. In a similar way, neither 2) nor 3) holds.

§25. (ℓ^2)-quasi-invariance

Th. 21.2 is generalized for non-stationary product measures as follows.

Theorem 25.1. Let μ be the measure given in (24.1).

1) If $\forall k$, $\frac{d}{dx}\sqrt{f_k(x)}\in L^2(dx)$ and $\sup_k \|\frac{d}{dx}\sqrt{f_k(x)}\|_{L^2}<\infty$, then we have $Y_\mu\supset(\ell^2)$.

2) Assume that $\forall k$, $x\sqrt{f_k(x)}\in L^2(dx)$ and $\sup_k \|x\sqrt{f_k(x)}\|_{L^2}<\infty$, then if and only if

$$\sum_{k=1}^{\infty}\left(\int_{-\infty}^{\infty} xf_k(x)dx\right)^2<\infty, \tag{25.1}$$

we have $R_\mu\supset(\ell^2)$ and the characteristic topology of μ is weaker than the topology of (ℓ^2).

Proof. Let $g_k(z)$ be the Fourier transform of $\sqrt{f_k(x)}$, then we have $\|\frac{d}{dx}\sqrt{f_k(x)}\|_{L^2} = \|zg_k(z)\|_{L^2}$. Since $Y_\mu^0=R_{\tilde\mu}$ and $\tilde\mu$ is the product measure of $|g_k(z_k)|^2dz_k$, 1) is derived from 2).

We have $R_\mu\supset(\ell^2)$, if and only if the characteristic topology of μ is weaker than the topology of (ℓ^2). This fact was proved in Th. 21.2, and the proof is valid also for non-stationary product measures.

Put $I_k=\int_{-\infty}^{\infty}x^2f_k(x)dx$ and $I=\sup_k I_k<\infty$. First assume that $c_k=\int_{-\infty}^{\infty}xf_k(x)dx=0$. The proof of Th. 21.2, 2), first half assures that

$$\left|\int\exp(i\sum_{k=1}^{\infty}y_kx_k)d\mu(x)-1\right|\le\prod_{k=1}^{\infty}\left(1+\frac{y_k^2}{\sqrt{2}}I_k\right)-1\le\exp\left(\frac{I}{\sqrt{2}}\sum_{k=1}^{\infty}y_k^2\right)-1.$$

Thus, the characteristic topology of μ is weaker than the topology of (ℓ^2).

If $c_k \neq 0$, then $\varphi_k(x)=f_k(x+c_k)$ satisfies the assumptions of 2), so that the characteristic function of $\prod_{k=1}^{\infty} \varphi_k(x_k)dx_k$, namely $\exp(-i\sum_{k=1}^{\infty} c_k y_k)\chi(y)$, is continuous in (ℓ^2). So $\chi(y)$ is continuous in the topology of (ℓ^2) if and only if $\sum_{k=1}^{\infty} c_k y_k$ is continuous in (ℓ^2), namely if and only if $(c_k)\in(\ell^2)$. (q.e.d.)

Remark 1. If $\mu \sim \mu'$, then we have $Y_\mu = Y_{\mu'}$ and $R_\mu = R_{\mu'}$. However even if μ satisfies the assumptions in Th. 25.1, μ' may not satisfy them.

For instance, if $\frac{d}{dx}\sqrt{f(x)} \in L^2(dx)$ and $\frac{d}{dx}\sqrt{f_1(x)} \notin L^2(dx)$, then we have $\frac{d}{dx}(\alpha_k\sqrt{f(x)} + \beta_k\sqrt{f_1(x)}) \notin L^2(dx)$ whenever $\beta_k \neq 0$. However, if $\{\beta_k\}$ decreases sufficiently rapidly to zero, the product measure of $(\alpha_k\sqrt{f(x_k)} + \beta_k\sqrt{f_1(x_k)})^2 dx_k$ is equivalent with the stationary product of $f(x)dx$. (α_k is chosen as to satisfy $\int_{-\infty}^{\infty}(\alpha_k\sqrt{f(x)} + \beta_k\sqrt{f_1(x)})^2 dx=1$).

We shall establish the converse of Th. 25.1 as a statement for the equivalence class of measures.

Theorem 25.2. Let μ be the measure given in (24.1).

1) If $Y_\mu \supset (\ell^2)$, then there exists a product measure $dM = \prod_{k=1}^{\infty} F_k(x_k)dx_k$ such that $\mu \sim M$, $\forall k$, $\frac{d}{dx}\sqrt{F_k(x)} \in L^2(dx)$ and $\sup_k \|\frac{d}{dx}\sqrt{F_k(x)}\|_{L^2} < \infty$.

2) If $R_\mu \supset (\ell^2)$, then there exists a product measure $dM = \prod_{k=1}^{\infty} F_k(x_k)dx_k$ such that $\mu \sim M$, $\forall k$, $x\sqrt{F_k(x)} \in L^2(dx)$ and $\sup_k \|x\sqrt{F_k(x)}\|_{L^2} < \infty$.

3) If $Y_\mu \supset (\ell^2)$ and $R_\mu \supset (\ell^2)$, then there exists a product measure $dM = \prod_{k=1}^{\infty} F_k(x_k)dx_k$ such that $\mu \sim M$, $\forall k$, $x\sqrt{F_k(x)} \in L^2(dx)$, $\frac{d}{dx}\sqrt{F_k(x)} \in L^2(dx)$, and $\sup_k \|x\sqrt{F_k(x)}\|_{L^2} < \infty$, $\sup_k \|\frac{d}{dx}\sqrt{F_k(x)}\|_{L^2} < \infty$.

Remark 2. In the case of 2) or 3), $c_k = \int_{-\infty}^{\infty} xF_k(x)dx$ should belong to (ℓ^2) in virtue of Th. 25.1.

Proof. First we shall prove 2). From Remark 3 of Th. 20.4, we have $\forall a = (a_k) \in (\ell^2)$, $\sum_{k=1}^{\infty} a_k^2 x_k^2 < \infty$ for μ-almost all x. Put

$$(25.2) \qquad dM = C\cdot \exp(-\sum_{k=1}^{\infty} a_k^2 x_k^2)d\mu,$$

where C is the normalization constant. Evidently $M \sim \mu$ and M is the product measure of

$$(25.3) \qquad F_k(x_k) = C_k \exp(-a_k^2 x_k^2) f_k(x_k),$$

where C_k is the normalization constant. Since $C = \prod_{k=1}^{\infty} C_k$, we have $\lim_{k\to\infty} C_k = 1$.

Now, choose (a_k) suitably as to satisfy $\sup_k I_k(a_k) < \infty$ where

$$(25.4) \qquad I_k(a) = \int_{-\infty}^{\infty} x^2 \exp(-a^2x^2) f_k(x)dx.$$

(Note that $\|x\sqrt{F_k(x)}\|_{L^2} = \sqrt{C_k I_k(a_k)}$ and $\lim_{k\to\infty} C_k = 1$). $I_k(a)$ is continuous and monotonically decreasing on $(0,\infty)$. We have $\lim_{a\to\infty} I_k(a) = 0$ and $\lim_{a\to 0} I_k(a) = \int_{-\infty}^{\infty} x^2 f_k(x)dx$ (=may be ∞).

From Remark 3 of Th. 20.4, the function

$$W(y) = \int_{\mathbb{R}^\infty} \exp(-\sum_{k=1}^{\infty} y_k^2 x_k^2)d\mu(x) = \prod_{k=1}^{\infty} \int_{-\infty}^{\infty} \exp(-y_k^2 x^2) f_k(x)dx$$

is continuous on (ℓ^2). So that we have

$$(25.5)' \qquad \exists\delta > 0,\ \sum_{k=1}^{\infty} y_k^2 \leqq \delta^2 \Rightarrow \sum_{k=1}^{\infty}\int_{-\infty}^{\infty} (1 - \exp(-y_k^2x^2))f_k(x)dx \leqq 1.$$

Since $te^{-t} < 1 - e^{-t}$ holds for $t > 0$, (25.5) implies

$$(25.6) \qquad \sum_{k=1}^{\infty} y_k^2 \leqq \delta^2 \Rightarrow \sum_{k=1}^{\infty} y_k^2\int_{-\infty}^{\infty} x^2\exp(-y_k^2x^2)f_k(x)dx < 1.$$

Now, if $I_k(0) \leqq \frac{1}{\delta^2}$, then put $a_k=0$. If $I_k(0) > \frac{1}{\delta^2}$, then define $a_k > 0$ by $I_k(a_k)=\frac{1}{\delta^2}$. We shall prove $(a_k)\in (\ell^2)$. We can show $\sum_{k=1}^{N} a_k^2 < \delta^2$ by mathematical induction as follows.

If $N=1$, applying (25.6) to $y=(\delta,0,0,\cdots)$, we have $\delta^2\int_{-\infty}^{\infty} x^2\exp(-\delta^2x^2)f_1(x)dx < 1$, so that $I_1(\delta) < \frac{1}{\delta^2}$. This means $a_1 < \delta$. Next, assume that $\sum_{k=1}^{N} a_k^2 < \delta^2$, and put $\alpha^2=\delta^2 - \sum_{k=1}^{N} a_k^2$. Applying (25.6) to $y=(a_1,a_2,\cdots,a_N,\alpha,0,0,\cdots)$, we get $\sum_{k=1}^{N} a_k^2I_k(a_k) + \alpha^2I_{N+1}(\alpha) < 1$. Since $I_k(a_k)=\frac{1}{\delta^2}$ for $a_k\neq 0$, we get $\alpha^2I_{N+1}(\alpha) < 1 - \frac{1}{\delta^2}\sum_{k=1}^{N} a_k^2=\frac{\alpha^2}{\delta^2}$, hence $I_{N+1}(\alpha) < \frac{1}{\delta^2}$, hence $a_{N+1} < \alpha$. Thus, we get $\sum_{k=1}^{N+1} a_k^2 < \delta^2$. This completes the proof of 2).

Next, we shall prove 1). Put

$$(25.7) \qquad F_k(x) = C_k\left[\frac{1}{\sqrt{2\pi}a_k}\exp(-\frac{x^2}{2a_k^2})*\sqrt{f_k(x)}\right]^2.$$

Then, the Fourier transform $G_k(z)$ of $\sqrt{F_k(x)}$ is given by $G_k(z)=\sqrt{C_k}\exp(-\frac{a_k^2z^2}{2})g_k(z)$. The adjoint measure $\tilde{M}$ of M is the product measure of $|G_k(z_k)|^2dz_k$. Since $R_{\tilde{\mu}}=Y_{\mu}^0\supset(\ell^2)$, for a suitable $(a_k) \in (\ell^2)$ we have $\tilde{M} \sim \tilde{\mu}$, $\sup_k \| z G_k(z) \|_{L^2} = \sup_k \| \frac{d}{dx} \sqrt{F_k(x)} \|_{L^2} < \infty$. Since

$$(25.8) \qquad \prod_{k=1}^{\infty} \langle |G_k(z)|, |g_k(z)| \rangle_{L^2} = \prod_{k=1}^{\infty} \langle G_k(z), g_k(z) \rangle_{L^2}$$

$$= \prod_{k=1}^{\infty} \langle \sqrt{F_k(x)}, \sqrt{\dot{f}_k(x)} \rangle_{L^2},$$

$\tilde{M} \sim \tilde{\mu}$ implies $M \sim \mu$. (The first equality in (25.8) is valid only if $G_k(z)\overline{g_k(z)} \geqq 0$, which is true in this case).

Finally, we shall prove 3). Put

$$(25.9) \qquad F_k(x) = C_k \left[\frac{1}{\sqrt{2\pi} a_k} \exp(-\frac{x^2}{2a_k^2}) * \sqrt{\varphi_k(x)} \right]^2,$$

$$(25.10) \qquad \varphi_k(x) = C_k' \exp(-b_k^2 x^2) f_k(x).$$

Since $R_\mu \supset (\ell^2)$, for some $(b_k) \in (\ell^2)$ we have $\mu \sim \mu_1$ $(d\mu_1 = \prod_{k=1}^{\infty} \varphi_k dx_k)$ and $\sup_k \| x \sqrt{\varphi_k(x)} \|_{L^2} < \infty$. Since $Y_\mu = Y_{\mu_1} \supset (\ell^2)$, for some $(a_k) \in (\ell^2)$ we have $\mu_1 \sim \mu \sim M$ $(dM = \prod_{k=1}^{\infty} F_k dx_k)$ and $\sup_k \| \frac{d}{dx} \sqrt{F_k(x)} \|_{L^2} < \infty$. Now we shall show that $\sup_k \| x \sqrt{F_k(x)} \|_{L^2} < \infty$. Since $\| f * g \|_{L^2} \leqq \| f \|_{L^2} \cdot \| g \|_{L^1}$ and $x(f*g) = (xf)*g + f*(xg)$, we have

$$\| x \sqrt{F_k(x)} \|_{L^2} \leqq \sqrt{C_k} \left\{ \| x \sqrt{\varphi_k(x)} \|_{L^2} + \| \frac{x}{\sqrt{2\pi} a_k} \exp(-\frac{x^2}{2a_k^2}) \|_{L^1} \right\}$$

$$= \sqrt{C_k} \left\{ \| x \sqrt{\varphi_k(x)} \|_{L^2} + \sqrt{\frac{2}{\pi}} a_k \right\}.$$

Noting that $\lim_{k \to \infty} C_k = 1$, $\lim_{k \to \infty} a_k = 0$ and that $\| x \sqrt{\varphi_k(x)} \|_{L^2}$ is bounded in k, we conclude that $\sup_k \| x \sqrt{F_k(x)} \|_{L^2} < \infty$.

(q.e.d.)

CHAPTER 5. $\mathbb{R}_0^\infty$-INVARIANT MEASURES ON $\mathbb{R}^\infty$

§26. Infinite dimensional Lebesgue measure

Let $\mathbb{R}^\infty$ be the infinite product of real lines, and $\mathbb{R}_0^\infty$ be the subspace of $\mathbb{R}^\infty$ given by $\mathbb{R}_0^\infty=\{(\xi_n);\ ^\exists N,\ \xi_n=0 \text{ for } n\geqq N\}$. Then $\mathbb{R}^\infty$ is the algebraical dual of $\mathbb{R}_0^\infty$, and $\mathcal{B}_{\mathbb{R}_0^\infty}$ is nothing but the weak Borel field of $\mathbb{R}^\infty$.

The purpose of Chapter 5 is to discuss $\mathbb{R}_0^\infty$-invariant, σ-finite weak Borel measures on $\mathbb{R}^\infty$.

<u>Theorem 26.1</u>. Let μ be a σ-finite weak Borel measure on $\mathbb{R}^\infty$. Then μ is $\mathbb{R}_0^\infty$-invariant if and only if

(26.1) $\quad {}^\forall n,\ {}^\exists\mu_n$: σ-finite weak Borel measure on $\mathbb{R}^\infty$,

$$\mu = m^n \times \mu_n,$$

where m^n is the n-dimensional Lebesgue measure on $\mathbb{R}^n$.

<u>Proof</u>. (26.1) is sufficient because: Since m^n is $\mathbb{R}^n$-invariant, μ is $\mathbb{R}^n\times\{0\}$-invariant for every n. So that μ is $\mathbb{R}_0^\infty$-invariant because of $\bigcup_{n=1}^{\infty}(\mathbb{R}^n\times\{0\})=\mathbb{R}_0^\infty$.

Conversely suppose that μ is $\mathbb{R}_0^\infty$-invariant. By the Corollary of Th.19.5, we have $\mu\sim m^n\times\mu_n'$ for some probability measure μ_n' on $\mathbb{R}^\infty$. So, by Radon-Nikodym's theorem we have

(26.2) $\quad {}^\exists f(x,y) > 0, \quad d\mu = f(x,y)dm^n d\mu_n'.$

Both μ and $m^n\times\mu_n'$ being $\mathbb{R}^n\times\{0\}$-invariant, we have

(26.3) $\quad {}^\forall r\in\mathbb{R}^n, \quad {}^{\forall'}(x,y),\ f(x-r,y) = f(x,y),$

($\forall'$ means "almost all"). Therefore we have

(26.4) $\quad {}^{\forall'}y,\ \ {}^{\forall'}x,\ \ {}^\forall r\in\mathbb{Q}^n,\ f(x-r,y) = f(x,y),$

where $\mathbb{Q}$ is the set of all rational numbers. Since m^n is $\mathbb{Q}^n$-ergodic, (26.4) implies that $\forall' y$, $f(x,y)=$ independent of x ($= g(y)$: Put). So putting $d\mu_n = g d\mu_n'$, we get $\mu = m^n \times \mu_n$. (q.e.d.)

<u>Corollary 1</u>. On the space $\mathbb{R}^\infty$, no finite weak Borel measure is $\mathbb{R}_0^\infty$-invariant.

<u>Corollary 2</u>. Let μ be an $\mathbb{R}_0^\infty$-invariant, σ-finite weak Borel measure on $\mathbb{R}^\infty$. Then for a Borel set B_n in $\mathbb{R}^n$, we have

$$(26.5) \qquad \begin{cases} m^n(B_n) = 0 \Rightarrow \mu(B_n \times \mathbb{R}^\infty) = 0, \\ m^n(B_n) > 0 \Rightarrow \mu(B_n \times \mathbb{R}^\infty) = \infty. \end{cases}$$

<u>Proof</u>. In (26.1), μ_n is also $\mathbb{R}_0^\infty$-invariant so that $\mu_n(\mathbb{R}^\infty)=\infty$, thus we get (26.5).

(26.5) shows that the projection on $\mathbb{R}^n$ is unique for any $\mathbb{R}_0^\infty$-invariant σ-finite measure μ on $\mathbb{R}^\infty$. But we have infinitely many $\mathbb{R}_0^\infty$-invariant σ-finite measures on $\mathbb{R}^\infty$. Typical ones among them, called the infinite dimensional Lebesgue measures, will be studied below.

We shall denote with m the one-dimensional Lebesgue measure on $\mathbb{R}$. Let $Y=(Y_k)$ be a sequence of Borel sets of $\mathbb{R}$ such that $0<m(Y_k)<\infty$. We shall define two Borel measures m_k and λ_k by

$$(26.6) \qquad m_k(B) = m(B)/m(Y_k), \ \lambda_k(B) = m_k(B \cap Y_k).$$

m_k is σ-finite, while λ_k is a probability measure on $\mathbb{R}$.

Consider the product measure

$$(26.7) \qquad \mu_Y^{(n)} = \prod_{k=1}^{n} m_k \times \prod_{k=n+1}^{\infty} \lambda_k .$$

$\mu_Y^{(n)}$ is σ-finite on $\mathbb{R}^\infty$.

Theorem 26.2. For every weak Borel set E of $\mathbb{R}^\infty$, put

$$(26.8) \qquad \mu_Y(E) = \lim_{n\to\infty} \mu_Y^{(n)}(E).$$

This limit always exists, and becomes a σ-finite $\mathbb{R}_0^\infty$-invariant measure on $\mathbb{R}^\infty$. μ_Y lies on

$$(26.9) \qquad L(Y) = \bigcup_{n=1}^{\infty} L_n(Y), \text{ where } L_n(Y) = \mathbb{R}^n \times \prod_{k=n+1}^{\infty} Y_k.$$

Definition 26.1. The measure μ_Y defined in Th.26.2 is called the (infinite dimensional) Lebesgue measure supported by $L(Y)$.

Proof of Th. First we shall show that for $n>n'$

$$(26.10) \qquad \mu_Y^{(n)}(E \cap L_{n'}(Y)) = \mu_Y^{(n')}(E).$$

Since $\mu_Y^{(n)} = \prod_{k=1}^{n'} m_k \times \prod_{k=n'+1}^{n} m_k \times \prod_{k=n+1}^{\infty} \lambda_k$ and $\mu_Y^{(n')} = \prod_{k=1}^{n'} m_k \times \prod_{k=n'+1}^{n} m_k|Y_k \times \prod_{k=n+1}^{\infty} \lambda_k$, we have $\mu_Y^{(n')} = \mu_Y^{(n)}|L_{n'}(Y)$, namely (26.10). Here $m_k|Y_k = \lambda_k$ means the restriction of m_k on Y_k.

From (26.10) we know that $\mu_Y^{(n)}(E)$ is increasing with respect to n, therefore the limit (26.8) exists though it may be infinity.

Next we shall prove that $\mu_Y(E)$ is σ-additive, namely supposing that $\{E_k\}$ is mutually disjoint, we shall show

$$(26.11) \qquad \mu_Y(\bigcup_{k=1}^{\infty} E_k) = \sum_{k=1}^{\infty} \mu_Y(E_k).$$

Put $\mu_Y^{(n)}(E_k) = \alpha_{nk}$, then the left hand side of (26.11) is equal with $\lim_{n\to\infty} \sum_{k=1}^{\infty} \alpha_{nk}$, while the right hand side is equal with $\sum_{k=1}^{\infty} \lim_{n\to\infty} \alpha_{nk}$. Since α_{nk} is monotonically increasing with respect to n, the sum and the limit commute mutually according to Lebesgue's theorem. Thus (26.11) has been proved.

Since $\mu_Y^{(n)}$ lies on $L_n(Y)$, the measure μ_Y evidently lies on $\bigcup_{n=1}^{\infty} L_n(Y)=L(Y)$. From (26.8) and (26.10), we have $\mu_Y=\mu_Y^{(n)}$ on the set $L_n(Y)$. Since $\mu_Y^{(n)}$ is σ-finite, μ_Y is σ-finite on $L_n(Y)$, therefore σ-finite on $\bigcup_{n=1}^{\infty} L_n(Y)=L(Y)$, hence σ-finite on $\mathbb{R}^\infty$.

Consider $Y^{(n)}=(Y_{k+n})_{k=1}^{\infty}$. Then we have

$$\mu_Y = \prod_{k=1}^{n} m_k \times \mu_{Y^{(n)}}, \tag{26.12}$$

because we have $\mu_Y=\lim_{n'\to\infty} (\prod_{k=1}^{n'} m_k\times \prod_{k=n'+1}^{\infty} \lambda_k)= \prod_{k=1}^{n} m_k\times\lim_{n'\to\infty} (\prod_{k=1}^{n'-n} m_{k+n}\times \prod_{k=n'-n+1}^{\infty} \lambda_{k+n})= \prod_{k=1}^{n} m_k\times\mu_{Y^{(n)}}$.

From (26.12) and Th.26.1, μ_Y is $\mathbb{R}_0^\infty$-invariant because $\prod_{k=1}^{n} m_k$ is a constant multiple of m^n. (q.e.d.)

Definition 26.2. Take $Y_k=[-\frac{1}{2},\frac{1}{2}]$ for every k. The corresponding μ_Y is called the standard Lebesgue measure on $\mathbb{R}^\infty$ and denoted with μ_∞.

The measure μ_∞ lies on $L=\bigcup_{n=1}^{\infty} (\mathbb{R}^n\times[-\frac{1}{2},\frac{1}{2}]^\infty)=\{(x_n);\ \exists N, |x_n|\leq\frac{1}{2}$ for $n\geq N\}$, therefore μ_∞ lies on (ℓ^∞).

§27. $\mathbb{R}_0^\infty$-ergodicity and mutual equivalence

We shall use the notations introduced in §26.

Theorem 27.1. Let μ be an $\mathbb{R}_0^\infty$-invariant, σ-finite weak Borel measure on $\mathbb{R}^\infty$. If μ satisfies

$$\mu(L(Y)^c) = 0 \text{ and} \tag{27.1}$$

$$\mu(L_0(Y)) < \infty \text{ where } L_0(Y) = \prod_{k=1}^{\infty} Y_k, \tag{27.2}$$

then we have $\mu=c\mu_Y$ with $c=\mu(L_0(Y))$.

Proof. First we shall apply Th.26.1. Since m^n is a constant multiple of $\prod_{k=1}^{n} m_k$, we get

(27.3) $\qquad \forall n, \exists \mu_n$: σ-finite weak Borel measure on $\mathbb{R}^\infty$,

$$\mu = \prod_{k=1}^{n} m_k \times \mu_n.$$

Now, for a weak Borel set E of $\mathbb{R}^\infty$, put

(27.4) $\qquad \mu^{(n)}(E) = \mu(E \cap L_n(Y)).$

Then (27.1) implies that $\mu(E)=\lim_{n\to\infty} \mu^{(n)}(E)$, thus it is sufficient to prove $\mu^{(n)}=c\mu_Y^{(n)}$ for each n.

Let $(B_k)_{k=1}^{n'}$ $(n'>n)$ be a sequence of Borel sets in $\mathbb{R}$. Then from (27.3) and (27.4) we have

(27.5) $$\mu^{(n)}\left(\prod_{k=1}^{n'} B_k \times \mathbb{R}^\infty\right) = \mu\left(\prod_{k=1}^{n} B_k \times \prod_{k=n+1}^{n'} (B_k \cap Y_k) \times \prod_{k=n'+1}^{\infty} Y_k\right)$$
$$= \prod_{k=1}^{n} m_k(B_k) \times \prod_{k=n+1}^{n'} \lambda_k(B_k) \times \mu_{n'}\left(\prod_{k=n'+1}^{\infty} Y_k\right).$$

Consider the special case of $B_k=Y_k$, then the left hand side of (27.5) becomes $\mu(L_0(Y))=c$, so that we get $\mu_{n'}\left(\prod_{k=n'+1}^{\infty} Y_k\right)=c$. Thus (27.5) implies that $\mu^{(n)}\left(\prod_{k=1}^{n'} B_k \times \mathbb{R}^\infty\right)=c\prod_{k=1}^{n} m_k(B_k) \times \prod_{k=n+1}^{n'} \lambda_k(B_k)$, hence $\mu^{(n)}=c\prod_{k=1}^{n} m_k \times \prod_{k=n+1}^{\infty} \lambda_k = c\mu_Y^{(n)}$. (q.e.d.)

<u>Corollary 1</u>. μ_Y is $\mathbb{R}_0^\infty$-ergodic.

<u>Proof</u>. Suppose that μ is $\mathbb{R}_0^\infty$-invariant and $\mu \leqq \mu_Y$, then μ satisfies (27.1) and (27.2), so we have $\mu=c\mu_Y$. This means that μ_Y is $\mathbb{R}_0^\infty$-ergodic (cf. Th.6.1'). (q.e.d.)

<u>Corollary 2</u>. Let μ be an $\mathbb{R}_0^\infty$-invariant and $\mathbb{R}_0^\infty$-ergodic, σ-finite weak Borel measure on $\mathbb{R}^\infty$. Then we have $\mu \sim \mu_Y$ if and only if $0<\mu(L_0(Y))<\infty$. Furthermore under this condition, we have $\mu=c\mu_Y$ with $c=\mu(L_0(Y))$.

Proof. Because of ergodicity, $\mu\sim\mu_Y$ implies $\mu=c\mu_Y$ for some $c>0$, hence $0<\mu(L_0(Y))<\infty$.

Conversely assume $0<\mu(L_0(Y))<\infty$. Since $L(Y)$ is $\mathbb{R}_0^\infty$-invariant and contains $L_0(Y)$, we have $\mu(L(Y)^c)=0$. So, by Th.27.1 we get $\mu=c\mu_Y$. (q.e.d.)

Theorem 27.2. Let $Z=(Z_k)_{k=1}^\infty$ be another sequence of Borel sets in $\mathbb{R}$. We have $\mu_Y\sim\mu_Z$ if and only if

$$(27.6)\qquad \sum_{k=1}^{\infty} m(Y_k\Delta Z_k)/m(Y_k) < \infty.$$

Under this condition, we have $\mu_Z=c\mu_Y$ with

$$(27.7)\qquad c = \prod_{k=1}^{\infty} m(Y_k)/m(Z_k).$$

Proof. By the Corollary 2 of Th.27.1, $\mu_Z\sim\mu_Y$ is equivalent with $0<\mu_Z(L_0(Y)) < \infty$. By the definition of μ_Z, we have

$$(27.8)\qquad \mu_Z(L_0(Y)) = \lim_{n\to\infty}\Big(\prod_{k=1}^{n} m(Y_k)/m(Z_k)\times \prod_{k=n+1}^{\infty} m(Y_k\cap Z_k)/m(Z_k)\Big).$$

So, $\mu_Z(L_0(Y)) > 0$ implies that $\exists n_0$, $k\geqq n_0 \Rightarrow m(Y_k\cap Z_k)>0$, and $\prod_{k=n_0}^{\infty} m(Y_k\cap Z_k)/m(Z_k)>0$. The latter implies $\sum_{k=n_0}^{\infty} m(Z_k\cap Y_k^c)/m(Z_k)<\infty$, so

$$(27.9)\qquad \sum_{k=1}^{\infty} m(Z_k\cap Y_k^c)/m(Z_k) < \infty.$$

Under the condition (27.9), we can rewrite (27.8) as

$$(27.8)'\qquad \mu_Z(L_0(Y)) = \prod_{k=1}^{n_0-1} m(Y_k)/m(Z_k)\times \prod_{k=n_0}^{\infty} m(Y_k\cap Z_k)/m(Z_k)$$
$$\times \lim_{n\to\infty} \prod_{k=n_0}^{n} m(Y_k)/m(Y_k\cap Z_k),$$

so that $0<\mu_Z(L_0(Y))<\infty$ is equivalent with $\prod_{k=n_0}^{\infty} m(Y_k\cap Z_k)/m(Y_k)>0$, hence with

(27.9)' $$\sum_{k=1}^{\infty} m(Y_k \cap Z_k^c)/m(Y_k) < \infty.$$

Thus $\mu_Z \sim \mu_Y$ is equivalent with the system of conditions (27.9) and (27.9)'. But these conditions imply $\lim_{k\to\infty} m(Y_k)/m(Z_k) = 1$, so the conditions (27.9) and (27.9)' are equivalent with (27.6). Thus $\mu_Z \sim \mu_Y$ occurs if and only if (27.6) is satisfied.

Under the condition (27.6), we have $\mu_Z = c\mu_Y$ with $c = \mu_Z(L_0(Y))$. But using (27.8) again, we get $\mu_Z(L_0(Y)) = \prod_{k=1}^{\infty} m(Y_k)/m(Z_k)$. (q.e.d.)

Remark. Consider the special case of $Z_k = Y_k + x_k$, then we have $\mu_Z = (\mu_Y)_x$. Since $m(Z_k) = m(Y_k)$ for every k, we have $c=1$, thus we see "Whenever $(\mu_Y)_x \sim \mu_Y$, we have $(\mu_Y)_x = \mu_Y$."

Theorem 27.3. For the standard Lebesgue measure μ_∞ on $\mathbb{R}^\infty$, we have $(\mu_\infty)_x \sim \mu_\infty$ if and only if $x \in (\ell^1)$.

Proof. By Th.27.2, $(\mu_\infty)_x \sim \mu_\infty$ is equivalent with

(27.10) $$\sum_{k=1}^{\infty} m([-\tfrac{1}{2},\tfrac{1}{2}] \Delta [-\tfrac{1}{2}+x_k, \tfrac{1}{2}+x_k]) < \infty.$$

But the left hand side of (27.10) is equal with $2\sum_{k=1}^{\infty} \mathrm{Min}(|x_k|,1)$, so (27.10) is equivalent with $\sum_{k=1}^{\infty} |x_k| < \infty$, hence with $x \in (\ell^1)$.

(q.e.d.)

Remark. μ_∞ is an (ℓ^1)-invariant measure which lies on (ℓ^∞). (cf. Example 17.1).

§28. Equivalent probability measure of product type

The infinite dimensional Lebesgue measure μ_Y is equivalent with some probability measure of product type. We shall start with a criterion on the equivalence in a general setting.

Theorem 28.1. Let $(X,\mathcal{B})$ be a measurable space, G be a group of measurable isomorphisms, and μ_1, μ_2 be two G-quasi-invariant and G-ergodic measures on $(X,\mathcal{B})$. Then we have $\mu_1 \sim \mu_2$ if and only if

$$(28.1) \qquad \exists B_0 \in \mathcal{B},\ \mu_1(B_0) > 0,\ \mu_1|B_0 \sim \mu_2|B_0.$$

<u>Proof</u>. Evidently $\mu_1 \sim \mu_2$ implies $\mu_1|B_0 \sim \mu_2|B_0$ for $\forall B_0 \in \mathcal{B}$.

Conversely assume that (28.1) is satisfied. Since both μ_1 and μ_2 are G-ergodic, there exist $\varphi_k \in G$ $(k=1,2,\cdots)$ such that $\bigcup_{k=1}^{\infty} \varphi_k(B_0)$ is thick both in μ_1 and μ_2. (cf. Th.6.1).

Now for $B \in \mathcal{B}$, $\mu_1(B)=0$ is equivalent with $\forall k$, $\mu_1(B \cap \varphi_k(B_0))$ $=0$, hence with $\forall k$, $\mu_1(\varphi_k^{-1}(B) \cap B_0)=0$. The same holds for the measure μ_2 also. Thus $\mu_1|B_0 \sim \mu_2|B_0$ implies $\mu_1 \sim \mu_2$ on X.

(q.e.d.)

<u>Theorem 28.2</u>. Let ν_k $(k=1,2,\cdots)$ be a probability Borel measure on $\mathbb{R}$. Then, the product measure $\nu = \prod_{k=1}^{\infty} \nu_k$ is equivalent with μ_Y, if and only if $\nu_k \sim m$ for each k and

$$(28.2) \qquad \prod_{k=1}^{\infty} \int_{Y_k} \sqrt{f_k(t)}\,dt/\sqrt{m(Y_k)} > 0,$$

where $f_k = d\nu_k/dm$. (As usual, dt stands for $dm(t)$).

<u>Proof</u>. Note that ν is $\mathbb{R}_0^{\infty}$-quasi-invariant if and only if $\nu_k \sim m$ for each k, and that ν is $\mathbb{R}_0^{\infty}$-ergodic under this condition.

Since $\mu_Y(L_0(Y))=1$, a criterion of $\mu_Y \sim \nu$ is given by $\mu_Y|L_0(Y) \sim \nu|L_0(Y)$. But we have $\mu_Y|L_0(Y) = \prod_{k=1}^{\infty} \lambda_k$ which is a probability measure. So by Corollary of Th.7.1, the condition for $\prod_{k=1}^{\infty} \lambda_k \sim \prod_{k=1}^{\infty} \nu_k|Y_k$ can be obtained in the form:

$$(28.3) \qquad \prod_{k=1}^{\infty} \int_{Y_k} \sqrt{\frac{d\nu_k}{d\lambda_k}}\, d\lambda_k > 0,$$

which is equivalent with (28.2). (q.e.d.)

Example 28.1. Consider $f_k(t)>0$ such that

$$(28.4) \qquad \int_{-\infty}^{\infty} f_k(t)dt = 1, \text{ and}$$

$$f_k(t) = c_k \quad \text{for} \quad t\in Y_k.$$

If $c_k m(Y_k)<1$, note that such a function $f_k(t)$ certainly exists. Then, the left hand side of (28.2) is equal with $\prod_{k=1}^{\infty} \sqrt{c_k m(Y_k)}$, so that whenever $\prod_{k=1}^{\infty} c_k m(Y_k)>0$, we have $\mu_Y \sim \prod_{k=1}^{\infty} \nu_k$ where $d\nu_k = f_k dm$.

Thus, we see that for any $Y=(Y_k)$, μ_Y is equivalent with some probability measure of product type.

§29. The converse problem of §28

In this section, supposing that ν_k is given, we shall try to find $Y=(Y_k)$ such that $\mu_Y \sim \nu$.

Let $d\nu_k = f_k dm$ $(f_k(t)>0)$ be a probability measure on $\mathbb{R}$.

Theorem 29.1. The product measure $\nu = \prod_{k=1}^{\infty} \nu_k$ has an equivalent $\mathbb{R}_0^{\infty}$-invariant measure, if and only if

(*) we can choose $\{c_k\}$ such that $\prod_{k=1}^{\infty} f_k(t_k)/c_k$ converges almost everywhere in ν.

Furthermore under this condition, we have $\nu \sim \mu_Y$ for some $Y=(Y_k)$.

Corollary. An $\mathbb{R}_0^{\infty}$-invariant measure μ is written as $\mu=\mu_Y$ for some $Y=(Y_k)$, if and only if μ is equivalent with a probability measure of product type.

Proof of Theorem. We shall divide the proof into two parts. First half is to show that if ν has an equivalent $\mathbb{R}_0^{\infty}$-invariant measure μ, then ν satisfies (*). Second half is to show that if (*) is satisfied, then we can construct $Y=(Y_k)$ such that $\nu\sim\mu_Y$.

First, assume that ν has an equivalent $\mathbb{R}_0^\infty$-invariant measure μ, and put $\nu=fd\mu$. Since μ is written as $\mu=m^n\times\mu_n$ by Th.26.1, we have

$$(29.1)\qquad f(x) = \prod_{k=1}^{n} f_k(x_k)\varphi_n(x_{n+1},x_{n+2},\cdots)$$

$$\text{where}\quad \prod_{k=1}^{\infty} d\nu_{n+k} = \varphi_n d\mu_n.$$

Taking the logarithm of the both hand sides, we have

$$(29.2)\qquad g(x) = \sum_{k=1}^{n} g_k(x_k)+\psi_n(x_{n+1},x_{n+2},\cdots),\ \text{where}$$

$$g(x) = \log f(x),\ g_k(x_k) = \log f_k(x_k)\ \text{and}$$

$$\psi_n(x_{n+1},\cdots) = \log\varphi_n(x_{n+1},\cdots).$$

Since the Borel field $\mathcal{B}$ of $\mathbb{R}^\infty$ is generated by $\bigcup_{n=1}^{\infty} p_n^{-1}(\mathcal{B}_n)$, where p_n is the projection: $\mathbb{R}^\infty\to\mathbb{R}^n$ and $\mathcal{B}_n$ is the Borel field of $\mathbb{R}^n$, we can find the sequence $\{h_n(x_1,x_2,\cdots,x_n)\}$ such that h_n converges to g in probability ν.

$$(29.3)\qquad g(x) = \lim_{n\to\infty} h_n(x_1,x_2,\cdots,x_n)\ \text{in probability.}$$

Combining this with (29.2), we have $h_n(x_1,x_2,\cdots x_n)-\sum_{k=1}^{n} g_k(x_k) \fallingdotseq \psi_n(x_{n+1},\cdots)$ in probability, so that the both hand sides should be approximately constant. Namely, for some constant a_n we have $\lim_{n\to\infty}\{h_n(x_1,x_2,\cdots,x_n)-\sum_{k=1}^{n} g_k(x_k)-a_n\}=0$ in probability, so that

$$(29.4)\qquad \lim_{n\to\infty}\Big(\sum_{k=1}^{n} g_k(x_k)+a_n\Big) = g(x)\quad\text{in probability.}$$

Putting $a_n - a_{n-1} = -b_n$, we get $\sum_{k=1}^{\infty} (g_k(x_k) - b_k) = g(x)$ in probability. But since $\{g_k(x_k) - b_k\}$ is a family of mutually independent random variables, its convergence in probability implies its almost everywhere convergence (Levy's theorem). So that $\sum_{k=1}^{\infty} (g_k(x_k) - b_k)$ converges almost everywhere, thus $\prod_{k=1}^{\infty} f_k(x_k)/c_k$ converges almost everywhere for $c_k = \exp b_k$.

Conversely assume that (*) is satisfied. We shall apply the following:

Kolmogorov's three series theorem. A series of mutually independent random variables $\sum_{n=1}^{\infty} X_n$ converges almost everywhere, if and only if the three series $\sum_{n=1}^{\infty} p_n$, $\sum_{n=1}^{\infty} m_n$ and $\sum_{n=1}^{\infty} v_n$ converge where p_n is the probability of $|X_n| > 1$, m_n and v_n are the mean and the variance of $\overline{X}_n$ which is defined by $\overline{X}_n = X_n$ for $|X_n| \leqq 1$ and $= 0$ for $|X_n| > 1$.

Applying this theorem to our case, we have $\sum_{k=1}^{\infty} p_k < \infty$ and $\sum_{k=1}^{\infty} v_k < \infty$, where

$$(29.5) \qquad p_k = \nu_k(Y_k^c), \quad Y_k = \{t_k;\ |g_k(t_k) - b_k| \leqq 1\}$$
$$= \{t_k;\ e^{-1} \leqq f_k(t_k)/c_k \leqq e\},$$

$$(29.6) \qquad v_k = \int_{Y_k} (g_k(t) - b_k')^2 d\nu_k(t),$$
$$b_k' = b_k + \int_{Y_k} (g_k(t) - b_k) d\nu_k.$$

Since we have $f_k(t) \geqq e^{-1} c_k$ on Y_k, we have $m(Y_k) < \infty$. From $\sum_{k=1}^{\infty} p_k < \infty$, we have $p_k < 1$ except finite numbers of k. Since $p_k < 1$ is equivalent with $m(Y_k) > 0$, we can conclude $0 < m(Y_k) < \infty$ except finite numbers of k. For the exceptional k, we shall choose arbitrary Y_k satisfying $0 < m(Y_k) < \infty$.

Now, we shall prove that $\nu \sim \mu_Y$. We shall prove

$$(29.7)\qquad \sum_{k=1}^{\infty}\left(1-\left(\int_{Y_k}\sqrt{f_k(t)}dt\right)^2/m(Y_k)\right) < \infty,$$

which is equivalent with (28.2). Since we have $\exists\alpha>0$, $|\log s|\leq 1 \Rightarrow |\log s|\geq\alpha|1-s|$ and since $|g_k(t)-b_k'|\leq 2$ for $t\in Y_k$, we have $|g_k(t)-b_k'|=|2\log\sqrt{c_k'/f_k(t)}|\geq 2\alpha|1-\sqrt{c_k'/f_k(t)}|$. Here $\log c_k'=b_k'$. Therefore from $\sum_{k=1}^{\infty} v_k<\infty$, we get

$$(29.8)\qquad \sum_{k=1}^{\infty}\int_{Y_k}(1-\sqrt{c_k'/f_k(t)})^2 f_k(t)dt < \infty,$$

hence

$$(29.9)\qquad \sum_{k=1}^{\infty}[\nu_k(Y_k)-2\sqrt{c_k'}\int_{Y_k}\sqrt{f_k(t)}dt+c_k'm(Y_k)] < \infty.$$

However, the summand is larger than $\nu_k(Y_k)-\left(\int_{Y_k}\sqrt{f_k(t)}dt\right)^2/m(Y_k)$, and $\sum_{k=1}^{\infty}p_k<\infty$ implies $\sum_{k=1}^{\infty}(1-\nu_k(Y_k))<\infty$, so that we get (29.7).

(q.e.d.)

Remark 1. A theorem equivalent with Th.29.1 was found by D. Hill. But his statement was more complicated. Later a simpler formulation with an elegant proof was presented by T. Hamachi. This lecture note adopts the latter formulation with a little modification.

Remark 2. As for Levy's theorem and Kolmogorov's three series theorem, confer with any book on the theory of probability, e.g. M. Loève, "Probability Theory". 4th edition (Springer) §17 and §18.

Theorem 29.2. The product of identical probability measures $\nu=\nu_1^{\infty}$ has no equivalent $\mathbb{R}_0^{\infty}$-invariant measure.

Proof. In this case, $f_k(t)$ in Th.29.1 does not depend on k. Then, (*) implies $f_1(t_k)/c_k\to 1$ almost everywhere, so that $f_1(t) =\lim_{k\to\infty} c_k$ in probability ν_1, thus $f_1(t)$=const. almost everywhere.

This is impossible because $d\nu_1 = f_1 dm$ is a probability measure.

(q.e.d.)

Theorem 29.3. Let $(\ ,\ )_1$ be an inner product on $\mathbb{R}_0^\infty$, and g_1 be the gaussian measure defined on $\mathbb{R}^\infty$ corresponding to $(\ ,\ )_1$. Let L $(\subset\mathbb{R}^\infty)$ be the topological dual space of $\mathbb{R}_0^\infty$ with respect to $(\ ,\ )_1$. If $\mathbb{R}_0^\infty$ is contained densely in L, then g_1 is $\mathbb{R}_0^\infty$-quasi-invariant and $\mathbb{R}_0^\infty$-ergodic. But g_1 has no equivalent $\mathbb{R}_0^\infty$-invariant measure.

Proof. Using the usual (ℓ^2)-inner product $(\ ,\ )$, we can write the given $(\ ,\ )_1$ as

$$(29.10) \qquad (\xi,\eta)_1 = (A\xi, A\eta) \quad \forall \xi,\eta\in\mathbb{R}_0^\infty$$

for some linear operator A from $\mathbb{R}_0^\infty$ onto $\mathbb{R}_0^\infty$. Then $L=A^*((\ell^2))$ and $g_1=\tau_A g$, where g is the gaussian measure corresponding to $(\ ,\ )$.

If $g_1\sim\mu$ and μ is $\mathbb{R}_0^\infty$-invariant, then $g\sim\tau_{A^{-1}}\mu$, and $\tau_{A^{-1}}\mu$ is $A^{*-1}(\mathbb{R}_0^\infty)$-invariant. Since g is a product of identical probability measures, if $\tau_{A^{-1}}\mu$ is $\mathbb{R}_0^\infty$-invariant, this contradicts to Th.29.2. So, the theorem is reduced to the following:

Lemma 29.1. Let Y and Y' be subspaces of $\mathbb{R}^\infty$. If μ is symmetric, Y-invariant, Y-ergodic and Y'-quasi-invariant, then μ is Y'-invariant. Here the symmetricity of μ means $\mu(-B)=\mu(B)$.

Proof. For $\forall x\in Y'$, since both μ and μ_x are Y-invariant and Y-ergodic, we have $\mu_x = c_x\mu$ for some constant $c_x>0$. Since we have $\check{\mu}=\mu$ (where $\check{\mu}(B)=\mu(-B)$), we get $c_x\mu=c_x\check{\mu}=(\mu_x)^\vee=\check{\mu}_{-x}=\mu_{-x}=c_{-x}\mu$, hence $c_x=c_{-x}$. This means $c_x=1$, because of $\mu=(\mu_x)_{-x}$.

Thus μ is Y'-invariant. (q.e.d.)

Remark. This lemma holds even if μ is quasi-symmetric, namely even if we assume $\check{\mu}\sim\mu$ instead of $\check{\mu}=\mu$. Because, $\check{\mu}\sim\mu$ implies $\check{\mu}=c\mu$, so combining with $(\check{\mu})^{\vee}=\mu$, we get $c=1$, thus the quasi-symmetricity of μ implies the symmetricity of μ.

§30. Linear transformation of μ_∞

The Lebesgue measure μ_Y is a type of $\mathbb{R}_0^\infty$-invariant σ-finite measures on $\mathbb{R}^\infty$. Another type of $\mathbb{R}_0^\infty$-invariant σ-finite measures is provided by the linear transformation of μ_∞ (more generally of μ_Y).

Let A be a linear operator from $\mathbb{R}_0^\infty$ onto $\mathbb{R}_0^\infty$. For a weak Borel set E of $\mathbb{R}^\infty$, put

$$\tau_A\mu(E) = \mu(A^{*-1}(E)), \tag{30.1}$$

where A^* is the adjoint operator of A. (A^* is a measurable mapping on $(\mathbb{R}^\infty, \mathcal{B}_{\mathbb{R}_0^\infty})$). Since $\tau_A\mu_\infty$ is $A^*((\ell^1))$-invariant, if $\mathbb{R}_0^\infty \subset A^*((\ell^1))$, then it is $\mathbb{R}_0^\infty$-invariant. Moreover if $\mathbb{R}_0^\infty$ is dense in $A^*((\ell^1))$ (or equivalently if $A^{*-1}(\mathbb{R}_0^\infty)$ is dense in (ℓ^1)), then $\tau_A\mu_\infty$ is $\mathbb{R}_0^\infty$-ergodic. Thus, we get infinitely many $\mathbb{R}_0^\infty$-invariant and $\mathbb{R}_0^\infty$-ergodic measures in this way.

For an element x of $\mathbb{R}^\infty$, the translated measure $(\tau_A\mu_\infty)_x$ (= hereafter we shall denote it with $\tau_x\tau_A\mu_\infty$) is again an $\mathbb{R}_0^\infty$-invariant and $\mathbb{R}_0^\infty$-ergodic measure. In general, the measure $\tau_x\tau_A\mu_\infty$ is not written in the form of μ_Y, so we obtain a new type of $\mathbb{R}_0^\infty$-invariant and $\mathbb{R}_0^\infty$-ergodic measures in this way.

As a special case, if every Y_k is a closed interval, namely if $Y_k=[a_k,b_k]$, then putting $x_k=(a_k+b_k)/2$ and A: $(\xi_k)\to$

$((b_k-a_k)\xi_k)$, we have $\mu_Y=\tau_x\tau_A\mu_\infty$. So such a μ_Y is a special case of $\tau_x\tau_A\mu_\infty$.

Example 30.1. The simplest example of $\tau_A\mu_\infty$, which is not equal with any μ_Y.

Let $\{e_k\}$ be the canonical base of $\mathbb{R}_0^\infty$, i.e. $e_k=(0,\cdots,0,1,0,\cdots)$ (only the k-th coordinate is 1), and define A by

$$\text{(30.2)} \qquad A:\ e_{2k-1} \to \frac{1}{\sqrt{2}}(e_{2k-1}+e_{2k})$$
$$e_{2k} \to \frac{1}{\sqrt{2}}(-e_{2k-1}+e_{2k}),\ k=1,2,\cdots.$$

Then we have $\tau_A\mu_\infty\perp\mu_\infty$. To prove this, we shall show $\tau_A\mu_\infty(L_0)=0$ where $L_0=[-\frac{1}{2},\frac{1}{2}]^\infty$.

Since $A^{*-1}(L_0)=\{x;\ |x_{2k-1}+x_{2k}|\leq\frac{\sqrt{2}}{2},\ |x_{2k-1}-x_{2k}|\leq\frac{\sqrt{2}}{2}$ for $k=1,2,\cdots\}$, we have

$$\text{(30.3)} \qquad A^{*-1}(L_0) = B_2^\infty,\ \text{where}\ B_2 = \{(x_1,x_2);|x_1+x_2|\leq\frac{\sqrt{2}}{2}\ \text{and}\ |x_1-x_2|\leq\frac{\sqrt{2}}{2}\}.$$

So that $\tau_A\mu_\infty(L_0)=\mu_\infty(B_2^\infty)=\{m^2(B_2\cap[-\frac{1}{2},\frac{1}{2}]^2)\}^\infty=\{2(\sqrt{2}-1)\}^\infty=0$.

More generally we can prove $\tau_A\mu_Y(L_0)=0$ for any $Y=(Y_k)$. Therefore we have $\tau_A\mu_Y\perp\mu_\infty$, so $\tau_{A^{-1}}\mu_\infty\perp\mu_Y$. Especially $\tau_{A^{-1}}\mu_\infty$ is not equal with any μ_Y.

Example 30.2. $\tau_A\mu_\infty$ is strictly $A^*((\ell^1))$-invariant and lies on $A^*((\ell^\infty))$. For instance, consider $A:\ (\xi_k)\to(\lambda_k\xi_k)\ (\lambda_k>0)$. If $\sum_{k=1}^{\infty}\lambda_k^{-2}<\infty$, then we have $(\ell^2)\subset A^*((\ell^1))$, thus we get an (ℓ^2)-invariant measure on $\mathbb{R}^\infty$.

In a similar way, choosing A suitably, for a given subspace X of $\mathbb{R}^\infty$, we can try to find an X-invariant measure on $\mathbb{R}^\infty$.

<u>Definition 30.1</u>. The following Λ_{μ_∞} is called the group of admissible linear transformations of μ_∞.

(30.4) $$\Lambda_{\mu_\infty} = \{A;\ \text{one-to-one and onto linear operator on } \mathbb{R}_0^\infty \text{ such that } \tau_A\mu_\infty \sim \mu_\infty\}.$$

<u>Theorem 30.1</u>. For a one-to-one and onto linear operator A on $\mathbb{R}_0^\infty$, we have $\tau_x\tau_A\mu_\infty \sim \tau_{x'}\tau_{A'}\mu_\infty$, if and only if

(30.5) $$AA'^{-1} \in \Lambda_{\mu_\infty} \quad \text{and} \quad x-x' \in A^*((\ell^1)).$$

<u>Proof</u>. First, note that in general we have $\tau_A\tau_B\mu = \tau_{BA}\mu$. Now $AA'^{-1} \in \Lambda_{\mu_\infty}$ implies $\tau_{A'^{-1}}\tau_A\mu_\infty \sim \mu_\infty$, so that $\tau_A\mu_\infty \sim \tau_{A'}\mu_\infty$. Since $\tau_A\mu_\infty$ is $A^*((\ell^1))$-invariant, $x-x' \in A^*((\ell^1))$ implies $\tau_{x-x'}\tau_A\mu_\infty \sim \tau_A\mu_\infty$, so that $\tau_x\tau_A\mu_\infty \sim \tau_{x'}\tau_A\mu_\infty \sim \tau_{x'}\tau_{A'}\mu_\infty$.

Conversely assume that $\tau_x\tau_A\mu_\infty \sim \tau_{x'}\tau_{A'}\mu_\infty$, then we have $\tau_{x-x'}\tau_A\mu_\infty \sim \tau_{A'}\mu_\infty$. Taking the reflection $\vee$ of both hand sides, we get $\tau_{x'-x}\tau_A\mu_\infty \sim \tau_{A'}\mu_\infty$. Thus we have $\tau_{x-x'}\tau_A\mu_\infty \sim \tau_{x'-x}\tau_A\mu_\infty$, so that $\tau_{2(x-x')}\tau_A\mu_\infty \sim \tau_A\mu_\infty$, hence $x-x' \in A^*((\ell^1))$, because of strict $A^*((\ell^1))$-invariance of $\tau_A\mu_\infty$. Now we get $\tau_A\mu_\infty \sim \tau_{A'}\mu_\infty$, so that $\tau_{AA'^{-1}}\mu_\infty \sim \mu_\infty$ hence $AA'^{-1} \in \Lambda_{\mu_\infty}$. (q.e.d.)

Since $\tau_A\mu_\infty$ is strictly $A^*((\ell^1))$-invariant, we have a simple necessary condition for $A \in \Lambda_{\mu_\infty}$: $A^*((\ell^1)) = (\ell^1)$.

We shall give some simple sufficient conditions.

<u>Theorem 30.2</u>. (1) $\Sigma \subset \Lambda_{\mu_\infty}$, where Σ is the group of all permutations $\sigma\colon (x_k) \to (x_{\sigma(k)})$.

(2) For a special case that $A\colon (\xi_k) \to (\lambda_k\xi_k)$ $(\lambda_k > 0)$, we have

(30.6) $$A \in \Lambda_{\mu_\infty} \Leftrightarrow \sum_{k=1}^{\infty} |1-\lambda_k| < \infty.$$

(3) If the image of $I-A$ is finite dimensional, and if

$A^*(\mathbb{R}_0^\infty) \subset (\ell^1)$, then $A \in \Lambda_{\mu_\infty}$.

Remark. $A\colon (\xi_k) \to (\xi_k + \sum_{j=1}^{\infty} \xi_j \delta_{1k})$ is such an example that the image of $I-A$ is one-dimensional, but $A^*\colon (x_k) \to (x_k + x_1)$ does not map $\mathbb{R}_0^\infty$ into (ℓ^1).

Proof of Theorem. (1) is evident, because $\sigma(L_0) = L_0$ (where $L_0 = [-\frac{1}{2}, \frac{1}{2}]^\infty$), hence $\tau_\sigma \mu_\infty = \mu_\infty$.

(2) Since $A \in \Lambda_{\mu_\infty}$ is equivalent with $A^{-1} \in \Lambda_{\mu_\infty}$, we shall find the condition for $A^{-1} \in \Lambda_{\mu_\infty}$. In this special case, putting $Z_k = [-\lambda_k/2, \lambda_k/2]$, we have $\tau_{A^{-1}} \mu_\infty = \mu_Z$. Applying Th.27.2, we have $A \in \Lambda_{\mu_\infty}$, if and only if $\sum_{k=1}^{\infty} m([-\frac{1}{2}, \frac{1}{2}] \Delta [-\lambda_k/2, \lambda_k/2]) < \infty$, which is equivalent with $\sum_{k=1}^{\infty} |1-\lambda_k| < \infty$.

(3) Assume that $A = I+T$, where $T(\mathbb{R}_0^\infty) \subset \mathbb{R}^n \times \{0\}$. Identify $\mathbb{R}^\infty$ with $\mathbb{R}^n \times \mathbb{R}^\infty$ and denote its element with (x,y) where $x \in \mathbb{R}^n$ and $y \in \mathbb{R}^\infty$. Then we have $T^* = 0$ on $\{0\} \times \mathbb{R}^\infty$, so $T^*(x,y) = T^*(x,0)$, which we shall denote with $(T_1^* x, T_2^* x) \in \mathbb{R}^n \times \mathbb{R}^\infty$. Thus $A^*(x,y) = (x,y) + (T_1^* x, T_2^* x) = (x + T_1^* x, y + T_2^* x)$.

Since μ_∞ is written as $\mu_\infty = m^n \times \mu_n$, we have

$$(30.7) \qquad \tau_A \mu_\infty(E) = \int_{\mathbb{R}^n} \mu_n(E(x) - T_2^* x)\, d(\tau_{A_1} m^n)(x),$$

where A_1 is the restriction of A on $\mathbb{R}^n \times \{0\}$. But since μ_n is (ℓ^1)-invariant and $T_2^* x \in (\ell^1)$ from the assumption, and since $\tau_{A_1} m^n \sim m^n$, we have the conclusion $\tau_A \mu_\infty \sim \mu_\infty$. (q.e.d.)

Remark. Th.30.2 (3) is valid for any $\mathbb{R}_0^\infty$-quasi-invariant σ-finite measure μ, if we replace the condition $A^*(\mathbb{R}_0^\infty) \subset (\ell^1)$ by $A^*(\mathbb{R}_0^\infty) \subset Y_\mu$, and the conclusion $A \in \Lambda_{\mu_\infty}$ by $A \in \Lambda_\mu$. The proof is carried out in the same way as above.

§31. Rotational invariance and ergodicity

A linear operator U from $\mathbb{R}_0^\infty$ onto $\mathbb{R}_0^\infty$ is said to be a rotation of $\mathbb{R}_0^\infty$, if $\|U\xi\|^2=\|\xi\|^2$ for $\forall\xi\in\mathbb{R}_0^\infty$, where $\|\xi\|=(\sum_{k=1}^{\infty}\xi_k^2)^{1/2}$ is the usual (ℓ^2)-norm. A rotation U is said to be finite dimensional, if there exists a finite dimensional subspace R such that $U=I$ on $R^\perp$.

Let G be the rotation group of $\mathbb{R}_0^\infty$, namely the group of all rotations of $\mathbb{R}_0^\infty$, and G_0 be the group of all finite dimensional rotations of $\mathbb{R}_0^\infty$.

Theorem 31.1. Every $\mathbb{R}_0^\infty$-invariant σ-finite measure μ on $\mathbb{R}^\infty$ is also G_0-invariant, i.e. $\forall U\in G_0$, $\mu=\tau_U\mu$.

Proof. Let $e_k=(0,\cdots,0,1,0,\cdots)$ (only the k-th coordinate is 1), and put

$$(31.1)\qquad G^{(n)}=\{U\in G;\ Ue_k=e_k \quad \text{for} \quad k>n\}.$$

Then we have $G_0=\bigcup_{n=1}^{\infty}G^{(n)}$. Since μ is written as $\mu=m^n\times\mu_n$ (by Th.26.1), and since m^n is rotationally invariant on $\mathbb{R}^n$, the measure μ is $G^{(n)}$-invariant for any n, hence μ is G_0-invariant. (q.e.d.)

Remark 1. Example 30.1 shows that μ_∞ is not G-invariant. It is an open question whether an $\mathbb{R}_0^\infty$-invariant and G-invariant measure exists or not.

Remark 2. In §11, we proved that every finite G_0-invariant measure can be expressed as a superposition of gaussian measures. This theorem fails for σ-finite G_0-invariant measure, because the infinite dimensional Lebesgue measure μ_Y is not equal (nor equivalent) with a gaussian measure.

Remark 3. Consider an operator A: $(\xi_k)\to(\lambda_k\xi_k)$ $(\lambda_k>0)$. If

$\sum_{k=1}^{\infty} \lambda_k^2 < \infty$, then we have $A^*((\ell^\infty)) \subset (\ell^2)$, so that $\tau_A \mu_\infty$ lies on (ℓ^2). Therefore, there exists a G_0-invariant non-Dirac measure on (ℓ^2).

Theorem 31.2. The standard Lebesgue measure μ_∞ is G_0-ergodic.

Proof. Assuming that

$$(31.2) \qquad {}^\forall U \in G_0, \quad \mu_\infty(E \Delta U^*(E)) = 0,$$

we shall prove that $\mu_\infty(E)=0$ or $\mu_\infty(E^c)=0$.

Let $\sum_0$ be the group of permutations of $\mathbb{N}$ such that $\sigma(k)=k$ except finite numbers of k. $\sum_0$ is regarded as a transformation group of $\mathbb{R}^\infty$ by σ: $(x_k) \in \mathbb{R}^\infty \to (x_{\sigma(k)}) \in \mathbb{R}^\infty$.

Since $\mu_\infty | L_0$ is the product of identical probability measures (= uniform probability measures on $[-\frac{1}{2}, \frac{1}{2}]$), it is $\sum_0$-ergodic. Namely, the condition

$$(31.3) \qquad {}^\forall \sigma \in \textstyle\sum_0, \quad \mu_\infty((E \Delta \sigma(E)) \cap L_0) = 0$$

implies $\mu_\infty(E \cap L_0)=0$ or $\mu_\infty(E^c \cap L_0)=0$.

Under the assumption (31.2), since (31.3) is satisfied, we have $\mu_\infty(E \cap L_0)=0$ or $\mu_\infty(E^c \cap L_0)=0$. Therefore, replacing E by E^c if necessary, it is sufficient to prove that $\mu_\infty(E \cap L_0)=0$ implies $\mu_\infty(E)=0$ under (31.2).

From $\mu_\infty(E \cap L_0)=0$ and (31.2), we see that

$$(31.4) \qquad {}^\forall U \in G_0, \quad \mu_\infty(U^*(E) \cap L_0) = 0$$

$$\text{hence} \quad \mu_\infty(E \cap U^{*-1}(L_0)) = 0.$$

So, it is sufficient to prove that

$$(31.5) \qquad {}^\exists U_n \in G_0,\ n=1,2,\cdots,\ \bigcup_{n=1}^{\infty} U_n^{*-1}(L_0) \text{ is thick in } \mu_\infty.$$

Now, since we have $\int_{-1/2}^{1/2} t^2dt=2\cdot\frac{1}{3}(\frac{1}{2})^3=\frac{1}{12}$, by the strong law of large numbers, we have

(31.6) $\lim_{N\to\infty}\frac{1}{N}\sum_{k=1}^{N}x_k^2=\frac{1}{12}$ for μ_∞-almost all $x=(x_k)$. In other words, putting

(31.7) $D = \{x\in\mathbb{R}^\infty ;\ \lim_{N\to\infty}\frac{1}{N}\sum_{k=1}^{N}x_k^2=\frac{1}{12}\}$,

we have $\mu_\infty(D^c)=0$. Since μ_∞ lies on $L_\infty=\bigcup_{n=1}^{\infty}(\mathbb{R}^n\times[-\frac{1}{2},\frac{1}{2}]^\infty)$, for the proof of (31.5) it is sufficient to show that

(31.8) $^\exists U_n\in G_0,\ n=1,2,\cdots,\ \bigcup_{n=1}^{\infty}U_n^{*-1}(L_0)\supset D\cap L_\infty$.

The group $G^{(n)}$ in (31.1) can be identified with the rotation group of $\mathbb{R}^n$, so it is separable in the usual topology (= Euclid topology of $\mathbb{R}^{n^2}$). Let $\{U_{kn}\}_{k=1,2,\cdots}$ be a dense subset of $G^{(n)}$. Then we shall show that

(31.9) $\bigcup_{n=1}^{\infty}\bigcup_{k=1}^{\infty}U_{kn}^{*-1}(L_0)\supset D\cap L_\infty$.

Take an arbitrary element x of $D\cap L_\infty$. Then, we have

(31.10) $^\exists N,\ \frac{1}{N}\sum_{k=1}^{N}x_k^2\leqq\frac{1}{9}$ and $|x_k|\leqq\frac{1}{2}$ for $k>N$.

There exists $U\in G^{(N)}$ such that $U^*x=(c,c,\cdots,c,x_{N+1},x_{N+2},\cdots)$ where $c=(\frac{1}{N}\sum_{k=1}^{N}x_k^2)^{1/2}\leqq\frac{1}{3}$. For this U, there exists U_{kN} such that $\|U-U_{kN}\|\leqq\frac{1}{2\sqrt{N}}$ where $\|\cdot\|$ means the operator norm on $\mathbb{R}^N$. Then we have $\|U^*x-U^*_{kN}x\|\leqq\frac{1}{2\sqrt{N}}(\sum_{k=1}^{N}x_k^2)^{1/2}\leqq\frac{1}{2\sqrt{N}}\cdot\frac{\sqrt{N}}{3}=\frac{1}{6}$, so putting $U^*_{kN}x=y=(y_1,y_2,\cdots,y_N,x_{N+1},x_{N+2},\cdots)$, we have $|y_k|\leqq c+\frac{1}{6}\leqq\frac{1}{2}$, thus $U^*_{kN}x\in L_0$. This implies $x\in U^{*-1}_{kN}(L_0)$, thus (31.9) has been proved.

(q.e.d.)

Remark. Th.31.2 is not valid for a general $\mathbb{R}_0^\infty$-invariant measure μ

For instance, assume that μ lies on (ℓ^2) (cf. Remark 3 of Th.31.1). Then, denoting with B the unit ball of (ℓ^2), we have $\exists n$, $\mu(nB)>0$. On the other hand, we have $\mu(nB^c)=\infty$, because $nB^c \supset \{x=(x_k);\ |x_1|>n\}$, thus μ is not G_0-ergodic.

§32. Invariance under homotheties

Using the standard Lebesgue measure μ_∞ on $\mathbb{R}^\infty$, we shall define a σ-additive measure $\overline{\mu}_\infty$ as follows:

$$\overline{\mu}_\infty(E) = \int_0^\infty \frac{1}{c}\tau_{cI}\mu_\infty(E)dc = \int_0^\infty \frac{1}{c}\mu_\infty(\tfrac{1}{c}E)dc. \tag{32.1}$$

The measure $\overline{\mu}_\infty$ is $\mathbb{R}_0^\infty$-invariant (actually (ℓ^1)-invariant), G_0-invariant, and invariant also under homotheties, i.e. we have

$$\forall \gamma > 0,\ \tau_{\gamma I}\overline{\mu}_\infty = \overline{\mu}_\infty. \tag{32.2}$$

In the case of finite dimensional space $\mathbb{R}^n$, the Lebesgue measure is a unique $\mathbb{R}^n$-invariant σ-finite measure, and it is not invariant under homotheties. In contrast with this, we have

<u>Theorem 32.1</u>. The measure $\overline{\mu}_\infty$ in (32.1) is a σ-finite weak Borel measure on $\mathbb{R}^\infty$. Therefore, on the space $\mathbb{R}^\infty$, there exists a σ-finite weak Borel measure which is invariant simultaneously under translations by the elements of $\mathbb{R}_0^\infty$ and under homotheties.

<u>Proof</u>. Since μ_∞ is σ-finite, there exists $\{E_k\}_{k=1}^\infty \subset \mathcal{B}$ such that $\mu_\infty(E_k)<\infty$ and $\bigcup_{k=1}^\infty E_k$ is thick in μ_∞. Here we can assume that $\{E_k\}$ is increasing and that each E_k is a subset of D defined in (31.7), because D is thick in μ_∞. Now, put

$$F_k = \bigcup_c \{cE_k;\ \tfrac{1}{k} \leq c \leq k\}. \tag{32.3}$$

Then F_k is the continuous image of $E_k \times [\frac{1}{k}, k]$ by the map

$(x,c)\to cx$ from $\mathbb{R}^\infty\times(0,\infty)$ onto $\mathbb{R}^\infty$. Since $\{cD\}_{c>0}$ is mutually disjoint and since $E_k\subset D$, this map is one-to-one on $E_k\times[\frac{1}{k},k]$, therefore F_k is a weak Borel set of $\mathbb{R}^\infty$ (cf. Th.14.3 of Part A).

Since $\bigcup_{k=1}^{\infty} E_k$ is thick in μ_∞, the set $\bigcup_{k=1}^{\infty} F_k$ is thick in every $\tau_{cI}\mu_\infty$, hence thick in $\overline{\mu}_\infty$. On the other hand, we have $\mu_\infty(c^{-1}F_k)=\mu_\infty(E_k)$ if $\frac{1}{k}\leq c\leq k$, $=0$ otherwise, so that we get

$$(32.4)\qquad \overline{\mu}_\infty(F_k) = \mu_\infty(E_k)\int_{\frac{1}{k}}^{k}\frac{dc}{c} < \infty.$$

This means that $\overline{\mu}_\infty$ is a σ-finite measure. (q.e.d.)

Remark. Though $\overline{\mu}_\infty$ is σ-finite, it is of $(0,\infty)$-type on the family of rectangles in $\mathbb{R}^\infty$. More precisely, we have the following statement.

Let A be a one-to-one linear operator from $\mathbb{R}_0^\infty$ onto $\mathbb{R}_0^\infty$, and assume that $A^*(\mathbb{R}_0^\infty)$ is contained densely in (ℓ^1). Then we have $\overline{\mu}_\infty(A^*(L_0)+x)=0$ or ∞. If it were positive and finite, the set of $c>0$ such that $0<\tau_{cI}\mu_\infty(A^*(L_0)+x)<\infty$ would have a positive measure, so especially for two different values c and c', we would have $0<\tau_{cI}\mu_\infty(A^*(L_0)+x)<\infty$ and $0<\tau_{c'I}\mu_\infty(A^*(L_0)+x)<\infty$. But from Corollary 2 of Th.27.1, this means that $\tau_{cI}\mu_\infty\sim\tau_x\tau_A\mu_\infty$ and $\tau_{c'I}\mu_\infty\sim\tau_x\tau_A\mu_\infty$, so that $\tau_{cI}\mu_\infty\sim\tau_{c'I}\mu_\infty$, which is a contradiction to $c\neq c'$.

Now, let μ be an arbitrary $\mathbb{R}_0^\infty$-invariant and $\mathbb{R}_0^\infty$-ergodic σ-finite measure on $\mathbb{R}^\infty$. We shall define $\overline{\mu}$ as follows:

$$(32.5)\qquad \overline{\mu}(E) = \int_0^\infty \frac{1}{c}\tau_{cI}\mu(E)dc = \int_0^\infty \frac{1}{c}\mu(\tfrac{1}{c}E)dc.$$

$\overline{\mu}$ is $\mathbb{R}_0^\infty$-invariant, G_0-invariant, and invariant also under homotheties. It is an open question whether $\overline{\mu}$ is always σ-finite or not, but if $\mu=\tau_A\mu_\infty$, then we have $\overline{\mu}=\tau_A\overline{\mu}_\infty$, thus $\overline{\mu}$ is σ-finite. Therefore, we conclude that on the space $\mathbb{R}^\infty$, there exist infinitely many σ-finite weak Borel measures which are invariant simultaneously under translations by the elements of $\mathbb{R}_0^\infty$ and under homotheties.

Theorem 32.2. 1) Let H be the transformation group of $\mathbb{R}^\infty$ which is generated by homotheties and translations by the elements of $\mathbb{R}_0^\infty$. Then, $\overline{\mu}$ is H-ergodic.

2) We have $\overline{\mu}\sim\overline{\mu}_\infty$, if and only if $\tau_{cI}\mu\sim\mu_\infty$ for some $c>0$. (This condition implies $\overline{\mu}=\alpha\overline{\mu}_\infty$ for some $\alpha>0$).

Corollary. $\Lambda_{\overline{\mu}_\infty}=\bigcup_c\{c\Lambda_{\mu_\infty};\ 0<c<\infty\}$, because $\tau_A\overline{\mu}_\infty\sim\overline{\mu}_\infty$ is equivalent with $\tau_{cA}\mu_\infty\sim\mu_\infty$ for some $c>0$.

Proof of 1). Assuming that

$$(32.6)\qquad \forall\xi\in\mathbb{R}_0^\infty,\ \overline{\mu}(E\Delta(E-\xi))=0,$$

$$\text{and}\quad \forall\gamma>0,\ \overline{\mu}(E\Delta(\gamma E))=0,$$

we shall prove that $\overline{\mu}(E)=0$ or $\overline{\mu}(E^c)=0$.

Since $\mathbb{R}_0^\infty$ is separable in the inductive limit topology of Euclid topologies on $\mathbb{R}^n\times\{0\}$, there exists a countable dense subset $X=\{\xi_n\}$. Then, every $\mathbb{R}_0^\infty$-invariant and $\mathbb{R}_0^\infty$-ergodic measure is also X-ergodic.

Now, from the first line of (32.6), we have $\forall n,\ \overline{\mu}(E\Delta(E-\xi_n))=0$, therefore $\forall' c, \forall n, \tau_{cI}\mu(E\Delta(E-\xi_n))=0$. Since $\tau_{cI}\mu$ is X-ergodic, this implies that

$$(32.7)\qquad \forall' c,\ \tau_{cI}\mu(E)=0\ \text{ or }\ \tau_{cI}\mu(E^c)=0.$$

From the second line of (32.6), we have $\forall\gamma>0$, $\forall' c$, $\tau_{cI}\mu(E\Delta(\gamma E))=0$, so that $\exists c_0$, $\forall'\gamma$, $\tau_{c_0I}\mu(E\Delta(\gamma E))=0$. This implies that $\forall'\gamma$, $\mu(\frac{1}{c_0}E)=\mu(\frac{\gamma}{c_0}E)$, so that $\mu(cE)$, hence $\tau_{cI}\mu(E)$ is constant for almost all c. Combining this with (32.7), we have $\forall'c$, $\tau_{cI}\mu(E)=0$ or $\forall'c$, $\tau_{cI}\mu(E^c)=0$. In the former case, we have $\bar{\mu}(E)=0$, while in the latter case, we have $\bar{\mu}(E^c)=0$.

Proof of 2). If $\mu\sim\mu_\infty$, then we have $\mu=\alpha\mu_\infty$ for some $\alpha>0$. In this case, from the definition (32.5), we have $\bar{\mu}=\alpha\bar{\mu}_\infty$. If $\tau_{cI}\mu\sim\mu_\infty$, then $\tau_{cI}\bar{\mu}=\alpha\bar{\mu}_\infty$, but $\bar{\mu}$ being invariant under homotheties, we have $\tau_{cI}\bar{\mu}=\bar{\mu}$, hence $\bar{\mu}=\alpha\bar{\mu}_\infty$.

Conversely, assume that $\bar{\mu}\sim\bar{\mu}_\infty$. Since the set D defined in (31.7) is thick in μ_∞, the set

$$F = \bigcup_c\{cD;\ 0 < c < \infty\} \tag{32.8}$$

is thick in $\bar{\mu}_\infty$, hence thick in $\bar{\mu}$. But we have $\gamma^{-1}F=F$ for $\forall\gamma>0$, so that F is thick in μ. Since μ is $\mathbb{R}_0^\infty$-ergodic and since each cD is $\mathbb{R}_0^\infty$-invariant, some c_0D should be thick in μ.

Now, $\mu_\infty(E)>0$ implies $\mu_\infty(E\cap D)>0$, so that the set $\bigcup_c\{c(E\cap D);\ 0<c<\infty\}$ has a positive $\bar{\mu}_\infty$-value, hence a positive $\bar{\mu}$-value, hence a positive μ-value. But since c_0D is thick in μ, we have $\mu(\bigcup_c\{c(E\cap D);\ 0<c<\infty\})=\mu(c_0E)$.

Thus $\mu_\infty(E)>0$ implies $\tau_{c_0^{-1}I}\mu(E)>0$. This means that $\mu_\infty\lesssim\tau_{c_0^{-1}I}\mu$, so that from $\mathbb{R}_0^\infty$-ergodicity of μ we get $\mu_\infty\sim\tau_{c_0^{-1}I}\mu$.

(q.e.d.)

NOTE

<u>Note for Chapter 1</u>. The theory of Haar measures and its converse problem, namely the theory of Weil topology, are classical results.

P.R. Halmos, "Measure Theory", van Nostrand 1950, Springer 1974.

A. Weil, "L'integration dans les groupes topologiques et ses application", Hermann, 1951.

In this lecture note, the author modified the definition of Haar measure, originally considered on a locally compact group, to include the case of a locally totally bounded thick group. By this modified definition, the theory of Weil topology becomes exactly the converse of the theory of Haar measures, because under Weil topology, the given group becomes only a thick group, not necessarily locally compact.

<u>Note for Chapter 2</u>. The properties of gaussian measures mentioned here are well known results. Cf. for instance, the books of Xia Dao-Xing or Skorohod stated below.

The result of §12 is so-called the theory of multiple Wiener integral, and is used as the base of further studies of functional analysis and probability theory, for instance Mariavin calculus.

<u>Note for Chapter 3</u>. Various results on quasi-invariance of infinite dimensional measures are found in:

Xia Dao-Xing, "Measure and integration theory on infinite dimensional spaces", Academic Press, 1972.

A.V. Skorohod, "Integration in Hilbert space", Springer 1974.

In this lecture note, the author discussed the conditions for quasi-invariance in terms of Kakutani topology and characteristic topology. It seems that the study of characteristic topology, or the topology of measure convergence, namely $L^0(\mu)$, is crucial for the criterion of quasi-invariance.

Note for Chapter 4. Main results come from

H. Shimomura, "An aspect of quasi-invariant measures on $\mathbb{R}^\infty$", Publ. RIMS, Kyoto Univ., vol. 11 (1976) pp. 749-773.

But the author improved his results and proofs, also adding some other results.

Note for Chapter 5. $\mathbb{R}_0^\infty$-invariant measures were firstly considered by C.C. Moore, and later studied by D. Hill.

C.C. Moore, "Invariant measures on product spaces", Proc. 5th Berkeley Symp. Math. Stat. & Prob. (Berkeley 1965) Vol.2 Part 2, pp. 447-459.

D. Hill, "σ-finite invariant measures on infinite product spaces", Trans. Amer. Math. Soc. Vol. 153 (1971), pp. 347-370.

But Hill's result as well as his discussion was very complicated. Still later, a simpler formulation with an elegant proof was given by T. Hamachi. In this lecture note, the author follows Hamachi's formulation.

T. Hamachi, "Equivalent measures on product spaces", Mem. of Fac. Sci. Kyushu Univ., Series A Vol. 27, No. 2 (1973) pp. 335-341.

The latter half of this chapter is based on the author's studies. Our future problem will be to find more general types of $\mathbb{R}_0^\infty$-invariant measures, and if possible, to determine all the $\mathbb{R}_0^\infty$-invariant measures on $\mathbb{R}^\infty$.

INDEX

absolutely continuous 114
adjoint measure 195
admissible linear transformation 242

Baire field 42
Baire measure 42
Baire set 42
Baire's theorem 183
Bochner's theorem 69
Borel field 38
Borel measure 38
Borel set 38
Borel's theorem 194
bounded set (of a locally
convex space) 95

Caratheodory's outer measure 9
characteristic function 69
characteristic topology 178
compact inner measure 23
compact regular measure 23
compact regular measurable space 35
compact space (set) 6
content 122
convexified Weil topology 139
convolution of measures 174
cotype 2 normed space 199

diameter of a set 38
directed set 32
direct sum (of locally
convex spaces) 92
dual topology 96

equi-measure cover 77
ergodic measure 145

finite algebra (of subsets) 9
finite dimensional rotation 244

gaussian measure 151

Haar measure 124
Hermite polynomial 173
Hilbertian semi-norm 77
Hilbertian type space (topology) 78
Hilbert-Schmidt (see HS)
homotheties invariant measure 247
Hopf's theorem 10
HS Hilbertian type topology 79
HS inner product 75
HS operator 72
HS semi-norm 77

image measure 17
image topology 35
inductive family (of
locally convex spaces) 93
inductive limit (of locally
convex spaces) 93
infinite dimensional
Laplacian operator 173
inner measure 22
inner regular measure 119
invariant measure
(under translations) 113
(under automorphisms) 144

Kakutani metric 189
Kakutani topology 190
Kolmogorov's extension theorem 39
Kolmogorov's three series theorem 237

Lebesgue measure
(infinite dimensional) 229
left Haar measure 125
left invariant measure 113
left quasi-invariant measure 113
Levy's theorem 237
linearized Kakutani topology 193
linearization of a topology 176
locally compact space 43
locally finite measure 129
locally precompact group 125
Luzin set (space) 51

measurable group 113
measurable isomorphism 14

measurable map 14
measurable space 14
measure (σ-additive or finitely additive) 9
measure convergence 178
measure space 17
metric of total variation 189
Minlos' theorem 73

n-dimensional width 102
nuclear semi-norm 99
nuclear space 85
(other definitions) 99, 102

outer measure 22
outer regular measure 119

positive definite function 69
precompact set 125
product measurable space 15
(uncountable case) 30
product measure 48
product topology 7
projective family
(of measurable spaces) 32
(of locally convex spaces) 93
projective limit measurable space
(countable) 16
(uncountable) 33
projective limit locally convex space 93
projective limit measure
(countable) 28
(uncountable) 33
projective sequence of measurable spaces 16

quasi-invariant measure
(under translations) 113
(under automorphisms) 144

right Haar measure 124
right invariant measure 113
right quasi-invariant measure 113
rotation 163
rotation group 163

Sazonov's theorem 77
Sazonov topology 80
self-consistent sequence (of measures) 18
separable space 38
separative measure 133
σ-algebra (of subsets) 9
σ-compact space 38
σ-generated measurable space 26
σ-precompact group 125
σ-separated measurable space 26
standard Lebesgue measure 230
standard measurable space 56
stationary product measure 208
strict inductive sequence 94
strong law of large numbers 166
strong topology 96
sum topology 35
Suslin set (space) 51
symmetric tensor 169
(representation) 171

thick subset 18
thick group 127
topological group 118
trace of a measure 18
transformed Lebesgue measure 240
Tychonov's theorem 7

U-separated set 120

Weil topology 134

Y-quasi-invariant 174